Welcome to

ONE WORLD UNITED
HEATHER LUTZ

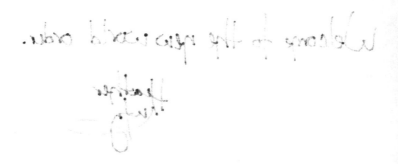

Copyright © 2024 by Heather Lutz
All rights reserved.

No part of this publication may be reproduced, distributed, or transmitted in any form or by any means, including photocopying, recording, or other electronic or mechanical methods, without the prior written permission of the publisher, except as permitted by U.S. copyright law. For permission requests, contact: Borderlands Media at borderlandsmedpub@gmail.com.

ISBN: 979-8-9898971-0-0 (paperback)
ISBN: 979-8-9898971-1-7 (e-pub)

The story, all names, characters, and incidents portrayed in this production are fictitious. No identification with actual persons (living or deceased), places, buildings, and products is intended or should be inferred.

Cover design and formatting by Borderlands Media.

First edition 2024

Published by Borderlands Media
275 W Farney LN
Las Cruces, NM 88005
https://www.borderlandsmedia.com/

To all the people who helped realize a dream come true.

CONTENTS

1. Chapter 1   1
2. Chapter 2   15
3. Chapter 3   29
4. Chapter 4   45
5. Chapter 5   61
6. Chapter 6   73
7. Chapter 7   83
8. Chapter 8   91
9. Chapter 9   115
10. Chapter 10   131
11. Chapter 11   141
12. Chapter 12   153
13. Chapter 13   161
14. Chapter 14   171
15. Chapter 15   181
16. Chapter 16   191

17.	Chapter 17	197
18.	Chapter 18	205
19.	Chapter 19	213

CHAPTER ONE

"Welcome to One World United."

The man who spoke knew the impact of his words. Conducted as to foster that ever-critical moment, a sea of murmurs would arise before him, words undulating in waves as the statement reached each individual. The ripples of sound led now to silence, for those who had spoken quietly watched the scene he had set, bound to unfurl just as quickly. No other statement was needed to start the buzz of the news feeds and the general public. Cameras clicked from every angle. The New World government was highly anticipated around the globe, leading supporters and non-supporters to gather for their first official council meeting.

Before the Great War, every country had their own way of running things. Each had their own government, religion and laws. Among them, there would always be one who thought themselves the best— in actuality, this mentality was shared by dozens. It fostered a constant rivalry, where there came a need to dominate those 'lesser' entities. People died of war, famine and sickness. Conquest escalated quickly. Yet, each time, it appeared the domineering countries would not expect their rivals to have ever more powerful allies, or for what seemed the world to unite against them. Empires always fell; it was what was 'right'... before the next would resurrect itself on remains of the fallen.

On and on; anyone looking back would know it had been this way since the beginning of written history. But, this war had set itself apart from those preceding. At its end, there were no governmental parties still standing. In its final breath stood a terrible consequence: total political annihilation. This brought the need to create something anew, to raise the white flag against eras without peace, and benefit the world at long last.

The speaker took a moment to allow the commotion to subside. He put his hands up in a gesture to settle, and the people obliged. Today, it seemed there were more civilian spectators than journalists, which surprised him, though it was not unwelcome. What astonished him much more was the lack of protesters present.

The people had been opposing the creation of a new government for some time. They wanted to govern themselves; the organization disagreed with this sentiment. Wasn't that how this had all begun in the first place? There wasn't much a single protestor could really do, not that it stopped groups from forming regardless. The council was worried about a resistance forming, but... after all the devastation? Few remained capable. The world was tired. For the moment, the majority were on their side, making it difficult for anyone to challenge.

ONE WORLD UNITED

Today marked the first meeting of the new 'regime', one given title, the "One World United". Their intent had been to televise it globally. Hence, the council invited as many still-living journalists as it could count. The idea was to be transparent. To Justin, however, this so-called 'transparency' was exactly why it was a farce. There were plenty of cards that the One World United, or the "OWU", as it was known, were holding close to their chest. Still... he could not deny the unique way they were going about this. Talk of the agenda surrounded the official launch of this new government, and more, was a promise of an announcement that would change the world forever. How much different could it be though?

The cameras from the news crews, the people doing podcasts, bloggers... vloggers, whatever else the world had created, were following behind the long line of people making their way inside the huge, wrought iron gates. The United Nations building had to be rebuilt from the ground up after the bombings, but they kept the original logo in its spot, where it had always been. The new one was remade from the old's remnants, as a reminder of how the organization progressed. The United Nations were the foundations from which the OWU was built, after all. It went from an organization made to protect and unify the world, to the ruling power of the world. They said it was to pay homage to the original plan of the UN, which was to achieve global peace and harmony.

Justin believed bringing it back like this was just proof of their dishonesty. A blinding, blaring siren of red-handed evidence; awareness of any part in centuries of history, whether by education or lived experience, would tell anyone who could listen that this would not change the trajectory of their world. Humanity would always be damned to repeat itself, so long as their nature stayed the same.

Behind dozens of spectators, the man followed silently. He had no desire to rush. Instead, he made a point to himself to blend into the crowd.

The marble floors were pristine with their silver and gold swirls. The regal palette was a theme throughout the entire vestibule. Justin couldn't help looking down at his faded brown penny loafers in contrast. The only flash of color below, until another person's foot entered the 'frame' of his vision. Stark. Looking up, further contrast was made between many bodies donning many colors, and the dull tones of the building. However dense this crowd was, the hall felt somehow desolate throughout most of its span. Decorum waited for its end, and only near there, glimpses of apparently randomly placed gemstones in the wall or floor emerged.

Felt tacky in a way– strange. Strategically eye-catching, maybe, if it could even be seen over the swaths of people. Some were more impressed than others, though Justin was not included in those 'some'. Regardless, their supposed superficiality aside, his gaze would follow the view elsewhere... eventually coming high enough to witness the wall of flags, the true 'crown gem' of the upper sector. Every country's symbol was displayed, each encased in their own frame. It was an undeniable masterpiece covering the furthest side of the room.

More so, it was a harrowing reminder of days behind them.

Columns and pillars held their old Greek influence, while the staircases looked like they had been taken directly from Buckingham Palace. In fact, unlike the logo, the OWU had not remade the United Nations building with its original pieces, but had taken fragments from the other governmental structures and buildings to create a *fortress*.

Echoing footsteps were mixed with the not-so-quiet whispers of appreciation as spectators walked inside. By the look of it, the entrance alone could accommodate the entire throng of people heading inside, as well as the neighboring village of over five hundred. Everyone stopped inside the atrium and waited eagerly for the next step.

"Behind me, you will be witness to the first ever public council meeting of One World United. We welcome you. I will beg that you be respectful and wait until the end of the meeting for any questions or comments. There will be a time when you will have three minutes to talk, if you wish." The announcer stopped a moment to look at each section of the crowd, akin to a father telling each of his children to behave.

"Also," he continued, while holding out a finger to signify a campaign-classic move: adding a rule. "I understand that a lot of people wish to record this piece of history on a variety of different types of devices, including cell phones. This is more than okay! In fact, we invite you to share this with the world and provide your own feedback. *But*, there is a caveat. You cannot take calls. You must put your phones on silent. It does not ring or otherwise disrupt this very important meeting. Agreed?" A smile graced the group in front of him, one that seemed so genuine that many of them smiled back without a worry.

Justin was surprised to find that even he was pleasant towards the older gentleman with salt and peppered hair, this host wearing a black suit with a red tie. While his attire was stereotypical of the event, his brown eyes were kind and reminded Justin oddly of a golden retriever. But he stopped smiling when the man moved out of the opening, and he realized that the moment he had been dreading was finally upon him.

"We have about ten minutes or so before we open the doors and let everyone into the conference room. This would be an ideal time to ensure that your phones are off. And…" There was a pause as people carried out the request. Not hindered, no, he hummed thoughtfully until the majority finished. "…Now, if anyone needs to freshen up, the bathrooms are down the hall and labeled accordingly. It could be a very long meeting, so I would suggest paying those rooms a visit at this time." He pointed near his left, to a narrow hall that had a sign engraved 'restrooms'.

As anticipated, people dispersed, following about the same rhythm that had been going since he began speaking. The man himself stood by and made small talk to the group that stayed before him, showing them the different parts of the vestibule and what they meant. Though, Justin wouldn't hear most of this, as he was among the former group moving about. He began to feel sick. Since he couldn't sit down, and he didn't have any want to talk to that 'golden retriever' of an announcer, he decided to see the restrooms and maybe splash some water on his face. He was aware that he looked grim and probably as pale and clammy as he felt, but he managed to force some semblance of calm by smiling slightly at everyone in his path.

The bathrooms were as outrageous as he expected them to be. Everything had been that marble engrained with gold. There were gold sinks, gold toilets. It was the most ostentatious thing he had ever seen. He thought he could now be called the expert of the subject since he was kneeling down in front of one of the auric thrones, about to throw up all the 'contents' of the food he had not eaten in days. It was strange to think how the body handled stress. There was nothing real to vomit, but that didn't mean his face hadn't been drenched in sweat, or that his body didn't retch all the bile that it could.

Despite the shock, it instantly stopped the process when he heard the door open. Standing up, Justin flushed the toilet and decided to walk out of the stall, where he'd make a beeline to the sinks. He'd splash water on his face, according to plan. At least once his face was dripping, no one would really know if he was weary, or just... wet.

Justin almost laughed out loud. He wanted to be out of his head so bad that he was having a conversation about the toilet and his bodily functions with himself. Every action or sight seen had been an internal monologue. Incredulously, he shook his head to himself in the reflection.

After a long moment of rehashing what he was even here for, he went to the towel dispenser and dried off before heading out of the restroom and back into the fray.

"Were you in any of the hot zones?" A voice behind him asked softly.

"Um... well, on the outskirts of one, yes," Somewhat slowly, empty, his voice came to him. "You?"

He remembered it like it was yesterday. How could he forget? The powers that be, those who were now named "One World

United", had found the brightest minds in the world and assembled them. Justin had been one of this group collected. On that morning, soldiers came to his house and banged on the door. His father knew it was coming, having worked with IT and coming across emails about what seemed to be called the 'gathering of the brains'.

"I was in Vancouver at the time. I missed it, but my extended family was not so lucky." The first man answered while he looked around for a seat.

He was just in time to hear the 'golden retriever' announce that they were to begin heading into the conference room.

"Ah, forgive the lack of seating...! We did not expect so many people to be here. However, after the leaders have settled into their places, you may find any seat that you wish, or stand anywhere you choose. Some may even wish to sit on the floor, which we would understand and be okay with. You have been standing outside in the elements most of the day and must be tired. We will begin in just a few more moments." Concluding disclosure, the 'gold' host again smiled, bowed to the gathering audience, and walked out of the room with his hands clasped behind his back.

The room was instantly a sea of wires that were being plugged into the walls for phones and cameras. People were lining the walls and finding places to stake claim. Chit-chat sprung up in the space and Justin started to feel panicked again. He had forgotten how much he hated crowds. Being in the open air at the gate had been easier than this. In that grand hall, however uncanny it felt, there had at least been a breeze... but this? *This* was a form of torture.

There was a stage behind the long table that had been laid out for the governmental peers. For a conference room it was big. Everything was exactly as it should have been but there was an oddity. Why would they need a stage? What could possibly be on the agenda that he didn't know about? But then again, perhaps— "Oof."

"Oh, excuse me! I'm Tyler from the New Time Magazine." The man that had backed into him held out a hand in a mix of apology and greeting, with a warm smile on his face.

"Hello I'm—" Oh god, he didn't want to use his real name. "I'm... Brandon. Just an interested party here to observe."

"Do you know what this announcement is going to be? I can't believe there haven't been any leaks on it yet."

"No idea. Maybe something needed to be added. You know, because 'officially' starting a whole new governmental system for the entire world isn't big enough," Justin replied with a tone more sarcastically than he meant to. "Whatever it is, it has to be pretty good for them to try and compete with the first meeting, is all I mean."

While a mistake at first, he did carry on with much of the same brashness. Maybe he should've cringed inwardly, stopped and tried to be nicer... Ah, well. The man didn't seem to mind much. His type must've heard from far worse, especially these days. Still, Justin knew well he was never good at friendly banter or small talk. It was why he hated people.

Anticipatedly, Tyler went on with his thoughts unbothered: "True. I just wonder, why a stage?" It was the same concern that Justin had, which he didn't really have an idea for either.

Before Justin could try and think of a response, the door by the stage opened. An immediate hush fell before the gathered crowd. The leaders of the new 'free' world came filing into the room, filling the chairs of the long table in the front of the room. The tapping of well-made black shoes signaled the approach of each, the steady *click, click, click...* White, leather-bound rolling chairs glided seamless by design, as not to disrupt or distract their entrance. Each settling motion revealed a glisten, that of the seats' gold threading... Showy. Of course it was. Gleaming just to draw in awe from the eyes of a deprived audience. Must've been an unspoken 'allotment of time', some means of justifying meeting in their *personal* theater.

Not to interrupt the speech nevertheless, no, never, as when the last leader sat down, the lights dimmed. Dark now, as to conceal the seamlines, yet kept bright enough as to see the smiles play across all the leaders faces and the reflections of the camera lenses.

"The first meeting of the One World United is officially in session. A few rules for our guests before we begin," The last person to join them was the first to speak. His voice boomed out confidently over the sea of people, almost as if he did not need to utilize any of the microphones arranged for them. But he was not idle. As he spoke, he walked through the group of people to the front of the room until he was in front of them. He looked like a professor addressing a class before an assembly.

"Foremost, though we are extremely happy to see all of you among us today, please understand that you are visitors and therefore should impose the same courtesy as you would as a guest in any one person's home. Do not interrupt our presentation. It took a long time to prepare this for you, and we would like to show it to you in its entirety before answering questions. As to say, there will be a time for questions, but only at the end. So, please wait until that time. We may answer something in the following segment that you are wanting to know. We recognize that this is our first public meeting and as such we expect it to be educational. Otherwise, non-inquiry based commentary is also to occur at the end; we expect this. And after all that, there will be a short– *I repeat, short-* and let me say one more time for the people in the back– *short-* meet and greet. This is not the time for an interview. This is for us to be able to thank you all for coming and see you briefly on a one to one basis. If you are going to take pictures, please make sure that there is no flash."

He gave the crowd a moment to do as instructed, as well as absorb the information that had been given. There was a rustle of people changing settings on their cameras and a few leaders at the table had poured themselves water.

"With that being said, hello! I am Grant Trailman, former President of the United States, and now one of the free leaders of One World United. I welcome you. This is the first meeting of One World United, and it is about time we begin proper." After his introduction, Mr. Trailman couldn't help himself grinning widely. Highlight of his career, wasn't it?

"With that, I am going to pass it over to Julia, who will take it from here!"

Julia was a lovely older woman, faint trails of laugh lines shown with a soft smile of the lips. Her long brown hair that glistened with silver streaks had been pulled back into a silver barrette, while her outfit had consisted of slacks and a button up shirt. Altogether, she looked very sleek and professional, her future words of utmost importance. Standing up, Julia had bowed slightly at the crowd before delivering her speech, a notable blue folder in hand.

"I know that some of you are apprehensive about the new world-wide government policies we are working on. I empathize. Today, I hope that I may help remedy those fears– and after so long of suffering, of such terrible hardships and losses, I *empathize*."

"Please, allow me to discuss what is in store for us all."

"Entering our new age together, there will never again be a great deficit, nor such disparity. No one state or country has more or less opportunity than the other. Monarchies, dictatorships, communism; too much power in any one sector is no longer to be the issue of any person on this planet. Slavery now is not only to remain outlawed where it has been, but to be recognized by every region of the world as a crime against humanity. Women are to have equal rights that are truly equal in every way possible, not to be encroached upon or removed conditionally. Those struggling through disasters of both war and nature– those without homes or livelihoods at all in wake of the war are offered panacea: to live in a world where human welfare is prioritized above all."

"In an oncoming era of peaceful unity, there is so much to be *hopeful* for. There is sanctuary that we, as people of all kinds, have so long wished for." She put her hands to her chest as though she was praying in thanks.

Her sermon continued. It grew more painstaking with every word that she spoke, in Justin's opinion. This 'promise' sounded more like a prelude to control that'd been offered like a loan unconditional– too good to be true. She was naming every injustice she could, and named it as good as solved.

How, exactly? What, using the same government as before but on a larger scale? Keeping the sorts of figureheads who let the damn war happen? Rename it, rewrap it, 'good enough'? This was just too much. He shifted in his seat uncomfortably. His palms were getting sweaty, like they always did when he was nervous.

"We have a long way to go," Julia spoke in a soft, modest tone before pausing for effect.

"It is only natural that it will take an adjustment period to get used to the new rules and regulations. We hope that everyone will help us transition smoothly and efficiently, though we are also confident that it will be met with positivity and obedience." She bowed to the crowd gracefully then retreated back to her chair.

Another man directly across from her stood up, a different folder in hand. This one had a business-like attire, better suited for a CEO of a bank, or an establishment manager elsewhere. With his pin striped suit and short, blond hair, combed and gelled stiffly, he looked like he had plenty of practice pinching numbers and firing people. There was a shuffling of pages, before he began.

"Like they said, we put together a small video." His voice was thick of an accent that Justin didn't recognize. "It is not a long one, but we are hoping it will help with some doubt as to why the One World United was created." He sat down and straightened the lapels of his suit, before nodding his head towards the technician, who'd already begun readying a projection screen.

The room went dark.

CHAPTER TWO

"Didn't know it would be like this," someone murmured in the darkness. It didn't sound as though they were trying to bring attention to themselves, but despite the population that had gathered, somehow they were the only one to speak. As though the crowd itself froze in the stillness.

After a moment of silence, the screen lit up with various images of chaos and destruction. Pictures of town after town having been ransacked right before battle. Many shops had broken windows, plundered of its goods as leftovers were discarded. Cars left on sides of the road were ablaze, a smoldering pile of metal in the background while people in gas masks wielded crude weapons. Nails hammered in bats, kitchen knives stained with blood, wooden planks hastily grabbed on hand.

"Someone help me, please!" The cry of a little girl echoed as a photo of a child was shown, her purple dress painted with red as she sat beside dead bodies. Presumably, her family. A group wielding similar weapons in hand could be seen nearby, but they weren't the focus right now. The cameraman had zoomed in on the bodies that surrounded the girl. They were badly beaten with bruises that coated any bare skin, blood dripping from agape mouths– it looked like their skulls were caved in.

"The devastation has reached all parts of the world. There is no place left that is unspoiled by death and war." A man with a turban and an accent spoke in front of a blazing mosque. Rubble was everywhere; shoes that had been discarded poked out of ash on the ground that blanketed the scenery like snow.

Image after image was displayed with one liners that broke the hearts of everyone in the room all over again. Everyone had their own story when it came to the Great War. Again, Justin was taken back to his own experience. He'd missed the worst of it, but not all. He remembered every detail all too well, burning in the back of his mind like the cars shown on screen.

It was Justin's father that woke him up before the sun rose, placing a finger to his lips with a shushing motion before urging him out of bed. Silently, he walked out of the room to meet with his mother, who had grabbed a backpack that had been packed for a few weeks. His father handed Justin his clothes and his shoes with urgency, while his mother walked towards the window to peek between the spaces of the blinds. They had been staying in a garage for weeks now, one that'd been owned by a friend.

Justin had never seen his parents so afraid. They were not fighters. His father was a small claims lawyer, and his mother worked at a daycare. They had decided to move to the bottom of Texas where Justin's dad had a compound. Neither of them were equipped to survive this warfare, nor did they have the experience. Almost everyone in the city that had something to lose had already left. The only people who stayed now were the ones destroying anything that could be called remnants. Homes, jobs, memories. Lives.

His father had received word through an IT email blast that he was given access for a certain court case. It stated that the government was to gather all potential brains, anyone well learned and willing, to work on a world-wide solution to this "problem". There was a list of names, and Justin's had been on it. His father started calling around last night, talking to his family about where to set up and buckle down for the fight ahead. They began moving from house to house, ending up here in this garage while waiting for a safe time to leave. Justin had feared for days that they were already too late.

That's when they heard a knock at the door.

"The governing power of Russia has mobilized units." A voice said, pulling Justin back to reality. The movie was now showing a brigade moving in unison against the poorly armed civilians. Then, it showed other governments moving against—you guessed it, more armies. They collided like waves against the shore, clashing with a ferocious intensity.

The scene panned, showing the United Nations building, then to a select few deliberating on how to solve their world problem.

Dates were written on the screen that Justin didn't pay attention to. It felt like half-drivel sympathy, understanding manufactured out of a pain that they likely didn't have to experience firsthand, a luxury he wasn't graced with. 'Trauma porn' for desiccated people, in simpler terms.

"The United Nations called on the leaders of the world. Every. Single. One. They all gathered and sought to find out how to end this global catastrophe. What did the people of the world want? The response was overwhelming, and yet so simple. Equality. Money. Freedom. Basic human rights. One by one, the chaos of the world was brought to order by the unity of the Nations. Creating what you see here– the One World United."

The images on the screen shifted from horror to hope, showing the leaders helping children out of bomb shelters.

"This is what will save us!" An elderly woman called out, clutching a bible to her chest.

"This is what should have always been- one people, the human race. No borders. No separation."

Finally, the snapshots were not of any one woeful person or government building or recordings of war, but of peace. Of a happy people. Different cultures bleeding into one, helping each other. The world was showing a happier, lighter version of itself- and still, Justin was not fooled. The music came to an end, and the lights went back to their normal settings.

"It isn't easy, trying to work through all the history and the emotions that we were raised to live in. We were taught to be separate from each other. We were not born believing we were the exact same as everyone else, with the same opportunities, same standings. And the divide got worse with time." Julia was speaking again. And this time it didn't just annoy Justin- it made his skin crawl.

"Either we strive to be better to each other, or we lose. Perhaps, now that there is balance in the world— true equality- we may be able to live better lives. Being alive is enough of a struggle without fighting each other along the way. Each of us now has a real chance at life, no matter where we're born." She finished the speech with a tear in her eye. The feeling seemed overwhelming for her, but Justin still wasn't buying it.

The former president began to speak after Julia, "The fact that people now have food and medicine... there is no more needless death due to famine, sickness, or war. People are living as they always should have. The vision of this being as it is right before us brings about many realizations that we had never even considered, because we never considered being where we are now. Our food distribution has tripled and will soon double— again! And, without the death rate that was needlessly rising every single day, we are feeling a little overcrowded. We have been trying to figure out how to co-exist with no qualms and without drying out the Earth's natural resources... which we're going to address with you, right now. With that, I will hand the spotlight over to our lead science guru. David? Take it away." Grant smiled, and sat down.

The 'science guru' David stood up slowly, studying the crowd. He was scrawny and pale, looking as though he'd had very few interactions in the daylight, but still— he stood with confidence and pride.

"Do we have any Star Trek fans out there?" David asked with a chuckle. A few people in the rooms nervously raised their hands, unsure of why this would even be a question. That television show had been canceled for so long that only a few people had even heard of it.

"I ask this because, in the beginning of most episodes from way back in the day, there is a quote that, up until recently, seemed highly unlikely. That quote is *'Space. The final frontier.'*" He used his best Picard impression, simpering all the while.

"I find this quote fascinating, because we can actually say this now. Our world is becoming too small. Quite frankly, we are running out of room. We have to find a way to branch out and learn how to live beyond that which we always have. Though-still, there are a few things to consider when we think about the fact that our world is now too small. For starters, we can't grow or expand the world itself. We are already building up and down instead of out, since we are running out of land. But, as humans, we also need to be able to have space for our children to be able to run and for people to have surface area to grow food and harvest. So the big question is- how do we fix this? Well– we scientists are always trying to build and create new things out of the things we already have. We like to discover and evolve. So what is something we have a ton of, that we have never really utilized or even really known what to do with?"

"*Space.*"

"Since we learned that we are not alone in the universe a few years ago, our own Space 9 division for Alien Representation and Communications has been in communication with those outside of ourselves, and found out new ways to create better space ports. This was a great discovery. Additionally, we have learned how to protect those on the space port from harmful UV rays while still allowing for sunlight to grace the entirety of the port- just as we experience the sun here. What this means–"

He smiled and took a deep breath,

"–is that we are finally ready to share: we have the capability to live in space as we would here on Earth! I have a few pictures for you on what the space port looks like, as well as what some of the life on that space port will entail." Excitedly, he sat, and nodded to the projection technician.

Their projection now showed some schematics and planning documents for a large, metal contraption. It looked intricate and detailed. The more the pictures panned out, the more it revealed that it was but a small piece of a *very* large space ship. It was domed with multiple layers. Some of it looked like they were large windows– other parts were open expanses. After four separate pictures, you could see the ship in its entirety. It was large enough to fit a small city. The presentation zoomed into an opening that leads you to look inside of the ship. A pool that looked to be Olympic sized; a football field, a school, training centers, shopping areas and even a market were inside. Multiple gardens... a movie theater. It was all there, and the only part of it that looked like it was military run were the housing units. These units were lined in halls, and were the equivalent of apartments. The ship had everything that a homely town would have, minus the people.

"So– the big news that everyone has been waiting on, is this: take this up and have it rest on top of the Earth's atmosphere. Then, move an entire town's worth of people there. Then, and only then, we will be able to see if we really have found a viable solution to our housing issues." The scientist explained.

There was a stir of murmuring and shifting in the crowd. A man that had been standing by a door nearby the stage watched as it opened beside him, the pause following alluded to this being a very important moment. Julia smiled and stood up with her hands raised to quiet the crowd.

"We actually have our lead family here, and they have agreed to meet with everyone. Is that okay?" She asked, her intention to incite response from those watching. She nodded to the door while smiling.

The rostrum door opened up and three people walked out onstage, looking like they won the lottery, genetic and otherwise. A woman walked out first, looking to be in her mid-to-late thirties. Her appearance was radiant– strong. Her brown hair swept the base of her neck and her bangs were a modern side sweep across her forehead. Her lips were full and only slightly glossed, and yet, she still had the appearance of being totally put together, all topped off with her tall, lean build. And, while she looked like she could handle any situation on her own, the man next to her looked like he could handle everyone *else* in the room, if need be.

The man held her hand in an almost protective way. He was tall and built in a way that screamed power. You knew he worked out, though it at least looked natural. His face was soft and loving as he looked at the woman beside him. And holding to his other fingers, was a hand belonging to a young boy. He was probably about ten or eleven, but he also had the same aura of the two adults– confidence and strength. He was definitely not shy. He grinned and raised his hand in greeting to everyone around him. It was not hard to see that this child was theirs. This was the starring family in this venture, and they looked the part.

"First, let me introduce you to our leading lady! This, here, is Alyssa Burns. She has been part of the space program since she was seventeen. We watched her grow and start a family. We have all been so excited to live vicariously through her and to see her progress in this life! She was our first choice for the mission, and she gladly accepted." Julia said with a smile. Alyssa, in turn, put her hand to her chest and bowed slightly.

"Her beautiful life would not be complete without her adamantly adoring husband, Sergeant James T. Burns. He was a Sergeant in the American Army and since his retirement there, has acted as Sergeant on our base– serving as a guard and personal trainer." He let go of his wife's hand briefly to give a wave, then grabbed on again.

"Together, this beautiful couple has a son named Steven. The space program and base has literally been a part of his entire life. And let's be honest, ladies and gentlemen, he is the true hero of this venture, right?" Julia beckoned and he bowed, playfully dramatic.

There was a feel of utter excitement in the air. Everyone was buzzed off the euphoria from the promise of such a bright, beautiful future. Alyssa looked around to see unabashed hope in the crowd. Everyone wanted to share in this joy. Everyone, except one. There was one she noticed in the crowd that did not share in the glee of the moment... He looked like he might be sick, actually. Alyssa had expected a mix of emotions, so having only one who was upset by this news was a really good turnout, in her book.

"We will open the floor for some questions now, to either us or the family on stage." A man said from the table with a heavy, slavic accent. He was older, but not bad looking. He wore a very large mustache though, and it was hard to make out his lips at all, Justin noticed.

Alyssa and James looked comfortingly at each other and sat on some chairs that had been brought out for them. The idea of their life being a sham and them not knowing it made Justin gag. He was unsure of how to handle the emotions, though. After all, he had also played his part in this.

"So... are you nervous about moving your family away from Earth?"

The question shook Justin out of his spiral and made him look up, listening intently. Perhaps, if he could just focus on them, he could make it through this. Every moment here was getting harder to stand.

"The way I see it," James started, "People move all the time. They move to different states, and different countries, and different areas all over the world. I'm not seeing it any differently than that."

"Steven, how do you feel about the mission?" A different reporter asked, with his recorder high in his hand.

"I have everything that matters here on this stage with me. I have my family. And you can make friends anywhere, if you try hard enough. Right, dad?" The little man looked at his father with the eyes of a boy looking at his hero.

"That's right, son." His father beamed down. Stupidly movie-esque, it felt too perfect.

"Are you at all worried that the fate of your lives rest entirely on a government that is newly formed with unknown flaws?"

Justin blurted out to the group. "Or how about– have you at all ever questioned if there are ulterior motives to this mission?" The words kept spilling, unable to stop himself. He thought hearing them speak about how willing they were about going would help ease him. Instead, it made him snap.

"Who is saying that?" James asked the group, standing up. He was almost menacing, though still calm.

"Why would you willingly risk your lives, and the life of your son for a mission you know nothing about?" There was a hint of hysteria in his voice as Justin continued.

"We believe in this mission, and have been a part of it from the beginning. We know everything we need to know about how safe it is, and as Captain of the ship, their lives are not in the hands of the government, but in mine. I am confident in my own capabilities." Alyssa spoke slowly, easily, reaching out and holding onto her husband's hand.

"Please contain yourself, sir, or you will be removed!" The Russian dignitary snarled out.

Desperate still, he almost pleaded: "You have to know that this mission is questionable at *best*. When did you decide that the life of your son meant so *little* that you would just blindly follow?"

He knew he wasn't getting through to them, not with their dedication. The right questions weren't being asked... but, it was like there had been a short in his brain. There were smart details that he couldn't bring out, because then the right people who happened to know who he was would come after him later. He was attempting caution even now as he tried to get them to listen at the same time, however, it only came off like a random citizen in a panic– it was becoming a jumbled mess.

This "One World United" wasn't as perfect as they made themselves out to be, they needed to know. But they wouldn't, would they?

There was a commotion in the back where security started to move quietly through the room to find the man speaking. There was just too much that Justin had held in for so long that he couldn't stop the flood gates opening. He had to try harder to warn them– he had to.

"Why don't the people of One World United explain to us why the *aliens* agreed to be so *helpful, free of charge*?" He managed to yell, nearly cut off as a guard grabbed him from the back of the arm, hauling him backwards.

Immediately the feel of the rooms shifted. The reporters stopped filming and looking at the family– started focusing on *him*. What else hadn't the organization shared? The room was in an uproar. Half of the media was following the man who was being removed, and the other half was trying to shout questions at both him and the government. The few who didn't fall into either of those categories stayed quiet, watching the show.

"What is your name?" One reporter called out.

"What secrets do you know about the One World United?" Another one yelled.

The man being forced out by two security guards knew he had mucked up his one chance to get to the family. It was obviously too late to try more. He had been more overcome with emotion than he could have thought possible. He turned to look behind him, at the chaos he was leaving in his wake. But instead of seeing the cameras glaring through a sea of outrage and confusion, his eyes caught the stare of Alyssa. For a brief moment, her questioning stare was louder than anything else in the room.

But the door slammed in his face and the world came back to him in a deafening roar. It was the end of a meager chance to make a difference, one that would resonate with him for years to come.

CHAPTER THREE

Jeff couldn't breathe. Even though he was outside, he felt like the air around him was as thick as molasses. He knew what it was: he was rattled. He had been to see Justin, who was still being held at the holding cell of One World United... and he couldn't believe what he had done. He knew he had to do some damage control, but he also knew there was no point in trying to get Justin out and back on base right now.

Before he knew where he was, Jeff was parking his vehicle at his apartment building. He'd been driving in a trance.

"*Ugh...* I'm going to be sick." He swayed as he got out of his truck. Jeff never did fly well, and just now he'd done a round trip in two days.

Walking up to his building gave him time to reflect on how this came to happen. He knew Justin had been struggling for a *while* now. Jeff understood what Justin was saying about not being able to let Alyssa lead her family on blind faith, sure... but now there was a much bigger, and far more immediate problem.

Jeff was an outcome engineer, though. He had already begun to formulate a way to spin this on its head. He'd dealt with the press before. There had to be a way to fix this. Rogue worker, maybe? No. That wouldn't work on the veteran journalists.

Maybe a stressed out worker who thought he was being righteous? Maybe he could say that there was too much PTSD from the wars. *That* could work. Not only was it true, but most of the time people were more forgiving when they knew you had been through a traumatic episode.

For Jeff, it was all about the science. He didn't understand this 'new nobility' of Justin's.

Jeff had been the youngest person to work on the "Space Living" situation. The technical term was really "Intergalactic Domestication Research Analyst"... but that was a mouthful. His sole purpose was to figure out how to successfully allow 'happy living' in space. Since science was beginning to change daily, people were having to adapt. Schools, religion, work, and lifestyles had all been shifted after the war, and now shifted again to the new science in their new world. It all was a domino effect where the biggest change and final domino falling would be that of this new space port.

The one thing that Justin had going for him was the controversy- not only on the new government, but on technology itself. With every new discovery, there was rebuttal; a conspiracy theory and a religious conflict. Every single time there was an advancement for the betterment of the people, there was a fight to make it "real" in the eyes of the world. But- there was a lot of positive PR with the space ports. There were a lot of people that knew that we couldn't go on the way that we were on Earth. So many thoughts were going around in Jeff's head- it was all so much to think about, to connect to Justin, how he felt, what he was going through- why he did what he did, and *whom* it was in front of.

Before he stepped into the vestibule, he took out his phone and a pack of smokes. He opened his Marlboros and put one in his mouth. Flicking his bic, he sucked on the stick until it was properly lit. Inhaled deeply. His phone had thirteen missed calls. Stared at it before dialing the number back, hitting call.

"I swear to God I am going to flip my shit and shoot you in the kneecaps. You went to see him. People *saw* you go to see him. What the good god damned hell were you thinking, Jeffrey?" Alex spat.

"Alex- first off- you don't know how to use a gun, let's not go there. And secondly, of course I went to see him. You would have been angrier if I hadn't gone to find out if he had started to blab to everyone in the cell he is in about our inner workings. You know as well as I do that Justin will talk to me before he talks to any of you. So," He stopped to take a drag, "you can thank me and ask me what he said instead of cursing me for doing what was right."

"Jesus, Jeffrey. Did he give you an actual excuse? Did he have an actual reason for this? Did he tell anyone things that he shouldn't have?" Alex's voice was on the verge of being manic. On the receiving end, he could really only bet he hadn't slept in days.

"Do you remember when we..." Jeff started on the phone.

"I really don't want to reminisce with you. Come on now. Just give me the deets. I need some sort of explanation here! I am trying to figure out how to save your job. Give me something I can use." That stopped Jeff. Alex was saving his job? Why was his job in danger?

"Alex, calm down! I did you a favor!" Jeff snapped, anger starting to boil. He shouldn't have to defend his work with anyone. But he also knew Alex. They had worked together for nearly ten years. There was no way Alex was going to see reason when he was like this. He tried again.

"Alex," Jeff spoke calmly, using his name repeatedly to break through the hysteria. It was a tactic he'd learned at some point- was supposed to be based in psychology. "Have you seen the news? Look at how many people have been looking up the space port recently. Then look at the comments." He paused and took another drag, his brows narrowing upon realizing his cigarette was almost out. "People are looking us up. They are concerned about overpopulation and they want this space port to work. They are behind us." A last huff, a long one, "I am not agreeing with what Justin did," He got the last sentence out quickly before Alex could get a word in, "But, *like I was telling you before*. Remember when we were working on that water treatment thing and everyone hated it until someone died of thirst and we could have helped? The bad PR turned and made us able to save half of the world's population. This is like that. Call me back after you look it up." Jeff hung up the phone

Somewhere in the last stretch of his tirade with Alex, he lost his smoke. He didn't see it on the ground though, so he must have had enough sense to put it out and toss it. That was his usual method anyway. Looked like he had to grab a new one. He gave Alex some good reasoning; now he just had to just wait. It could be minutes, or hours, but at least it was a start to fix the problem that Justin created. He struck the lighter with his thumb and the same moment that the fire was created, an extremely loud, obnoxious tone blared- his phone ringing. A passerby in front of him gave him a look like he had just slapped everyone within earshot.

He had his phone this loud and *terrible* because he slept like the dead, but it incited panic in him now. He dropped his bic, and his cigarette followed.

"Hallo!" Jeff spoke, almost dropping the phone while trying to get the cigarette before it rolled.

"Alright." Alex sighed, now a little calmer. "I mean, you're right. Our media coverage just went nuclear. Everyone wants to know more. There is a call for more interviews. More coverage. I mean- this is insane!"

"Yeah, I know." Which was a lil. He was only hoping.

"The new Oprah is coming out of retirement to do an interview with the family. Isn't that crazy?" Alex was still in awe.

"Isn't she in her nineties?" Jeff mused.

"Look man, I have to ask you something." Alex took a serious tone and paused to make sure Jeff was listening properly.

Jeff lit another and inhaled hot nicotine. He waited- he knew better than to prompt Alex.

"Did you know he felt like this?" Alex asked.

"Alex, are you asking if I knew he wanted to sabotage what we have been working on since I was out of college diapers?" Jeff asked, a little taken aback.

"Look Jeff, you guys were real close. I need to know if you feel the same way. Are you planning to pick up where he left off? I need to know where you are at, my man." The sincerity of his voice is what made Jeff answer truthfully, and calmly. He could understand feeling... worried.

"Alex," Inhale, exhale; fumes pressed through the teeth. Its slow burn provided the extra few seconds to put his thoughts together. "You know I have continually had my reservations about this whole deal. However, I refuse to stop my work because

I feel a twinge of nobility. If I need help sleeping at night, I will take a pill. We often progress so much in science, so quickly at times, that you either ride the wave with it to see where it takes you, or you miss out on what could be the adventure of a lifetime. I don't know if what we are doing is right. I don't like it. But I'm on this bull until the 8 seconds is up."

The truth was he really didn't understand why Jeff was so worked up. Working with the aliens is weird, but it wasn't like he, or any of them were really selling their souls. If anything, the human populace was getting more out of this than the aliens were. That had been his major reservation. But as time moved on, it was clear that there was no debt owed. This was just a spread of knowledge. And Jeff could live with that.

"Okay. I'll take that." Alex replied.

"So. We good?" Jeff asked.

"Just one more thing we have to talk about."

"Yeah?"

Alex was struggling to keep the cool that he had just gained, "Is he going to talk? What if someone tries to interview him?"

"Alex, Justin has been off base like, five times in ten years. We move him into a house on base, put him on house arrest, and then what is he going to do? He has never spoken out or lashed out before. You can't keep him in jail, but you can't let him out. So you give him small things to take care of and have him on house arrest. He's not going to talk to anyone. I am surprised his anxiety let him that far into the fray the first time." Jeff knew he was right, and though he hadn't considered it before it came out of his mouth, he instantly knew it was the right call.

"Hmm. I think that could be dangerous. But I see where you are coming from. When are you due back on base?"

"Tomorrow night. I am going to pack the rest of my stuff here, and the movers come in the morning. I'm tied to this mission for the duration, so I doubt I'll be here enough to keep it. So, I am letting it go. But you owe me dinner."

"You know the Board is going to want to talk to you about this when you return, right?" Noted, as if it just occurred to him.

"Meh. I've dealt with them before." Jeff rolled his eyes. He couldn't stand the Board. A bunch of rich old people who think they know the science and the ins and outs better than him. "No one is going to remember this incident. Not after the launch. And we are right on schedule."

"Okay. Yeah. You're right. I think." Alex sounded like he had deflated the remainder of his energy.

"Get some sleep. I'll make it right. I always do." Jeff assured him one final time before hanging up the phone.

He took a deep breath. *Fuck.* He hated the idea of moving on base. He had always had a space for himself besides that of the government. Now, he was giving it all up to be fully in the trenches for this project. There was nothing that made him an individual beyond One World United- he was just another of their workers. He knew it had to be done, but it didn't mean he liked it. Was this part of what drove Justin to what he'd done? His eyebrows furrowed while free fingers fussed with his belt loop.

"Whelp," He shrugged, tossing his smoke in the bin before going inside.

Before he even made it inside he felt anxiety on the horizon. He hated the idea of losing his apartment. It was his first big spend as an adult, when he moved out of his parents' house. Even though the location hadn't exactly been convenient, he had kept the apartment throughout college, grad school, the NASA program, and all that time in-between. He felt as though he was surrendering the final piece of himself before totally immersing into something that would never really see or treat him as a person. Sure- because the mission was unprecedented, he knew that the logical thing would be to live on base full time- be available at any time. But there was a minimum of ten years left on his contract with this mission... so he knew the logical thing was to say goodbye to this place. Still, it was just hard. There wasn't a way to logic himself out of his feelings, and he knew that, but he wished he could.

The climb up the stairs never seemed so short before. He had been thinking so hard on his next move that he didn't even get to savor the trip. He considered going back down and doing it all over again, but his knee gave a twinge. He was never very athletic.

Once inside he looked around at the display of half-filled boxes and open cabinets. His bachelor pad had been completed only about 3 years ago, with a sound system that could make the Star Wars movies feel like he was in a theater. He needed a drink. It would be a long time before he could have a night to just drink and do his own thing, so he wanted to have his last 'hoorah'. Searching the kitchen, he found only one cabinet had been left for packing, and it had all of his "nice" glasses and plates. He took a champagne flute out and cracked open a bottle of wine. A... *healthy* serving was poured, and he downed it. Then he filled another. He finished a bottle of wine while he packed the rest of the living room and kitchen. He finished another two bottles while battling the rest of the house. When everything was done, he could hardly stand without weaving.

"You aren't as smart as you thought you were." He mumbled to himself. He sat down right where he was, in the middle of the living room. There wasn't a need to walk right now. Nap. He needed to nap. Jeff didn't remember laying down, or closing his eyes.

Knock, knock, knock.

"What is it?" Jeff asked with a start. There he was- on the floor, slouched over in a terribly uncomfortable position. His neck hurt. And ouch, so did his head.

"We're here with the moving company?" A person from the other side of the door called back.

"Ugh. What time is it? I thought you weren't coming until eleven?" Jeff replied. He didn't know where he had put his phone. His watch was... nope. Not on his wrist. Huh. He hadn't remembered taking it off. Maybe he packed it for some reason.

"It's ah, yeah. It's ten 'till." They answered.

"Oh. Okay." He tried to get up. Failed. Tried again, and managed to stand up this time with only a slight tip. He staggered to the door and opened it. When he did, "Yeah. So. Everything needs to go. The bedroom set is still put together. I don't have tools to take it apart." Jeff managed, letting the group of men walk in.

"We have tools. We can do that for you, sir." The voice, or the man really, replied. He had a face now.

"I feel like I should take a shower. Is that weird?" He asked the man, but before there was a response, a phone rang.

"Is that yours, sir?"

"Umm. Yeah. But I don't know where it is ringing from."

Flustered, Jeff walked towards the sound and found his phone in the utensil drawer. The ringing stopped before he could answer.

"Well, shit." Jeff said a little more loudly than he anticipated. "...Ookay. So I am going to call this guy back and I am gonna go down and have a smoke while I do it. Unless you need me for anything?"

"No, sir." The familiar voice replied again. "We got this."

That was all he needed to hear- he grabbed his pack and walked out of the apartment. He had wanted to say something ultra-cool for some reason, but what would he say? 'I work for the government so if you steal my stuff I'll come find you'? He didn't even have his own tools or know how to get to where he was going. And duh- *they* were hired by the government. Instead, he just popped back in a second, grabbed his jacket, because it had his sunglasses in it, and walked back out.

He waited until he was all the way down the stairs before calling the number back. For some reason the stairway was a black hole for phones. Always had been. He called the number and sighed.

"What the hell is wrong with you? Going to see Justin right after all the shit he pulled at the OWU? What were you thinking? Are you on your way here? Because I need to kick your ass." The caller growled at him on the second ring, not even saying hello.

"Good morning to you too, *Ben*. I'm good, thanks for asking. I'll be there in a bit. The moving guys are here. I was up all night packing and, *uh-* drinking. So I'm a little bit hungover this morning," More sarcastically, he'd add, "But I'm so glad you called to check up on me. Very kind of you, you know, voicing your concerns."

"You fucked up, Jeff. Big time."

"Nah. I talked to Alex. We have things situated." He settled himself on top of the steps leading down to the street and pulled out a cigarette. "I know everyone is stressed out, but did you see the media coverage? I think Justin did us a favor. One nut job isn't going to break this mission, Ben."

"Well, look." Ben said, a little calmer, "I have your house ready for you to move in. Which, I guess, was the original reason for my call."

Jeff wondered why anyone ever doubted him to begin with, when he always managed to soothe them with something as simple as logic. Regardless... He glanced back over his shoulder, where there was an expected amount of heaving and shuffling. "Okay. I'm going to be following the movers. I guess they should know where we're going. In fact, they are carrying my couches out as we speak. So I think I'm going to smoke one more in peace, and then help them carry the rest of my stuff." Quipped, eyes now trailing after the sound, watching as four of the movers emerged, carrying both his couches to a truck up front of the building.

Jeff stretched out after he hung up the phone. The full feeling of the drunken sleep he had was rolling through him like waves lapping slowly up on a beach. He was used to working with no sleep. But this drunk... sleep... *thing* was worse. He stared up at the clouds for a minute, letting the breeze cool the sweat off his brow.

"Sir?" A voice called to Jeff, startling him awake. "You okay?"

"Huh?" Jeff scrambled to sit up.

"I asked if you were okay."

God, this looked lame. So much for being cool and impressionable. "Shit- how long was I out? On the steps... of my apartment building... like a total bum?"

"About an hour. We figured we should let you sleep a bit, but we are done now. And we didn't want to leave you. We have orders that you are going to follow us in." The man helped Jeff up off the floor and reached down to retrieve the phone that fell from his chest.

"Oh. Okay. Well, thanks. I need to get my keys and my bag from the apartment and I'll be right down. Is that okay?"

"Yes sir. We put your bag by the door, and laid your keys out on top of them. We didn't want to take them down in case there was anything else you wanted to do while you were up there."

Jeff huffed as he hurried up the steps. "Thank you again."

There was no time to soak in the final ascent of this home, no time to reminisce. He had to hurry. After everything, he did feel a bit better. Once Jeff grabbed the keys, retrieving the ones that went to the apartment, he was out the door again. Heading towards the fire safety box, he fumbled his hands behind and found the spot where he had placed a key in one of those... magnet key holder things. He entered his code, retrieved his spare, and took both items inside. The magnet box was placed within his bag, and the man took a moment to look around.

He went through every single room in the house. Above anything, he wanted to be sure he didn't leave something behind. Once every room, cupboard and door was checked, he rechecked for good measure- then reached behind the furnace and grabbed a thing kept in a fireproof box, as normal people did.

"Ooh- ooh, ah! Ouch!" Hurt about the normal amount- a lot. The box was quickly plopped into the sink. He gave it a moment as it sizzled, then grabbed it again. When he was sure it wouldn't burn a hole in his jacket, he put the box in it and left the keys in on the counter. The door to his apartment was closed, and he left that part of his life behind for good.

He silently thanked the apartment as he walked back down. Recalled all the times he had come home in different states of being. Drunk, stressed, sad, happy. He had so many memories there. Not all of them were good, but not all of them were bad, either.

He signed his leave agreement and gave it to the person at the front desk, then left.

"Alright then." He told the movers that were waiting for him by the truck. "Let's roll out."

"Just follow us, sir. We delivered your appliances earlier today, and took the liberty of hooking them up for you."

"Okay then. Well, lead the way." Jeff said as he headed towards his ugly red truck.

The ride was long. Too long. So. Very. Long. Jeff needed to sleep. He followed the truck that had his entire life in the back of it, and when the sun began its descent, he finally saw the entrance into the blockade complete with soldiers. He was back on base.

He watched as he saw the truck stop and talk to someone. Then they were waved on through, and so was Jeff. After about twenty more minutes of slow driving through the area, he pulled up to what was his new home. His house numbers were ironic to him. 5150. In medical terms this is the term used for an involuntary psych hold. The memory of his mother telling him about a song that had those numbers in it and what it meant made him smile. He would need to remember to call his mom eventually and tell her all about his new space. He would be forever living in an involuntary psych hold.

The movers got all of his stuff inside and put together his bed in less than two hours. He leaned out the back door and smoked a cigarette, looking forward to a long shower and sleep. His phone rang.

"Hello." He said, grimacing, after seeing who was on the caller ID.

"Hey, man, where you at?" Alex said on the line. He sounded a lot better than he had yesterday.

Stifling a yawn as he spoke, "Just got finished with the movers at my new casa. It's nice."

"Welp. No time for a welcome home mat. You're needed. Now. There are directions being sent to your email right now. I'll fill you in when you get here."

"What? No. Alex, I just got here. I smell like ass. Do I at least have time for a shower?"

"Negative, ghost rider. All hands are coming on deck. Right now."

"Fine. Okay. At least let me find the new keys to lock up this place. I'll be there as soon as I can."

It was a rare occasion for Jeff to smoke in his vehicle, but this was one of those moments. He was starting to feel the hangover bit of things again, and he desperately needed coffee. When he saw on the map that he was getting close, he stubbed out his cigarette and replaced it with two pieces of gum, followed with spraying himself and the car with the Febreeze he kept in his glove compartment. He went into his jacket, and found the box from behind the furnace. Shit. He took it out and put it in the compartment with the Febreeze. A hand back into his jacket for some Visine... Hopefully the eye drops would do their part.

Now, after some time driving, Jeff was a scientist, so he knew that there was no exact cause and effect of the cosmos. Time and space worked as it did without care for little things like humans. Emotions weren't anywhere in it. So, his truck didn't overheat and leave him almost to his destination just because he broke his own rule- smoking in the truck especially when he had no idea where he was going.

But overheat it did, and he was pretty sure it was leaking some sort of important liquid.

After parking on the side of the road he locked up his truck and walked all the way to the main entrance. He didn't have his new credentials to get into the building any other way. He stepped through the door and heard a welcoming *"blrrp"* sound the doors had always made.

"Hey, Jeff!" Alex called out from a high balcony, "Welcome home, my friend."

CHAPTER FOUR

"**M**om, we're on TV again!" Steven said with lazy enthusiasm, cheerios shoved into his mouth immediately after. "Mom? Mom!" He yelled for her again, even louder.

"Yes?" Alyssa peeked around the kitchen wall from where she was tending to their laundry.

"I don't think I ever asked, but is there going to be television up in space?"

"We will have television, yes. And a movie theater. But I don't think we'll have satellites. We're working on that though, so maybe in a few years we will..." His mother answered, folding a towel.

"I was curious about something else, too. I mean, it doesn't really matter, but it just doesn't make sense in my head." Steven pulled up a chair and sat himself down. There was a serious look on his face, so Alyssa placed down the towel she had in her hand and looked at him intently.

"Go on."

"Well, it's just... Why leave on a Wednesday?"

Oh? Was that it? The amusement in his expression made Alyssa chuckle. "You know what, buddy? I don't know." She laughed and picked up another towel to fold while giving more of an explanation, "I mean, I know there are people coming in from all over, and then there's a list of things to do. I think they just scheduled it for after those things would be done."

"What things?" Steven asked, raising a brow at her.

"I know on Monday, all those important people are coming in. Tuesday we go to the space center– though we do have to go to medical one more time on Earth-"

"Moooommmm-!" The boy whined at the mention of medical.

"We have to go back to the medical bay one more time on Earth, and then the next day we go." Alyssa spoke a little more sternly this time, displeased with the attitude.

"Okay, but... Mom?" After a small pause, Steven looked a little meek.

"Yes?"

"Why do we have to go to the medical bay so much? They always use needles." Though his brows furrowed in discomfort, he was trying really hard to not sound nervous.

"Well," Alyssa answered slowly. She always tried to answer any questions put to her by anyone to the best of her ability. "I think we have to go so much because we are going to be gone for so long. They have to have all of our information, as much as they can have, in case there is anything that comes up in the future. They need to know as much as they can to try to help."

"Like– but why?"

"In case we get super sick, or get hurt. We may not have what we need there, so they may have to send something to us. But that way, they have everything we need."

Her continuation was met with a groan from Steven. "But then we have to go to medical when we land...!"

"Honey, I don't know what to tell you. We have to do what we have to do. None of us like having to go through that. And there are people younger and older than you that are doing the same thing. We all have to do it."

"But mom-!"

"Are you done eating?" Alyssa promptly cut off the fit.

"Yes." A grumpy answer.

"Yes, what?"

"Yes ma'am."

"Good. Go play until dad comes home. Then we're gonna eat, shower, and go to bed." With that, she stood up to take the towels to the bathroom.

Steven turned around to leave, but he had only taken one step before his mom grabbed him from behind and pulled him into a big bear hug.

"Ah!"

They stayed like that for a minute. Neither liked the irritation in the other's voice when they were talking, and as his mother, she did have to remedy that.

"I'm sorry bud. I don't like it either. But we will get through it all, together. I love you, okay?" She held onto him.

"I love you too, mom." He turned in her arms and wrapped his own around her properly.

Alyssa quietly stood and watched Steven go to his room. She began to think about the conversation with her son and how strange life was, recalling a talk with her father about the future. This wasn't the future she saw for herself though.

"Alyssa!" A man, her father, yelled at her. "If you lose your footing you're going to fall!"

They were at a rock climbing place that was in the city. He and Alyssa would go once a month, and it was mostly amazing. 'Mostly', because the remainder was terrible at the same time. Her father was the Captain of a unit that did nothing but take on combat missions. It was scary, hard work, something she saw little of and he made sure that she hadn't. Despite this, he was keen on his special permission to take his daughter along with him wherever they went.

Once a month, no matter what happened, they would leave base and go rock climbing. After that, they would grab pizza and talk, before heading back. There, they'd resume silent cohabitation. All in all, the time out wasn't as terrible as it could have been. It was mostly terrible for Alyssa because she could neither compete with nor keep up with her father. It was a continual disappointment. Why couldn't they have changed things up? He was a stubborn man, so they only ever went here on their occasions– to a rock climbing place.

She hadn't been thinking about the activity this trip. Not as clearly as she needed to. She lost her footing, just as her father said she would, and down she came– there were, of course, the protective lines. They prevented her falling in a crunch, and instead, gently lowered her to the ground.

"Where are you today, Alyssa?" Her dad looked down at her with his hands on his hips, noticing a sense of distraction from her. He wasn't angry, he was genuine.

"I don't... I don't really want to do this right now, actually. I want to talk about mom today." She said quietly.

He sighed and got on his knees before her, looking eye to eye as he spoke, "Well... I take you here for a reason, we could start there. This is what she and I used to do before we had you, so I make sure to keep it alive. Mom is always with us."

"...Mom did this?" There were new tears swelling in her eyes as Alyssa spoke up.

"Yeah. Once a month. That was our date for the longest time before we got married. We kept it up long after, that is until..."

He didn't need to finish the sentence. She died, that's what he meant, and Alyssa understood without needing him to speak the words. But the thought that her mother and father had done this very thing made her want to try again, to follow in the footsteps left behind. She would reach the top this time. Like her mother.

"Okay. One more time, then pizza?" Alyssa asked her father, standing up and brushing off her hands on her pants.

"If you're sure." He never pushed her. If she said she wanted to leave, they would be gone immediately. But if she wanted to stay, then stay they would.

"I'm sure. I need to, now." She was determined, well on her way to start at the bottom. Somehow, her father always knew what to say. While Alyssa was climbing, sweating and making decisions on where to put her hand or foot next, her father was a narrator in the background.

"See, your mom said she wanted to try it once. And I have always been a bit extreme. Back then they used to have passes to get in, so I had bought a monthly pass for us as our first date. Then I ended up getting a year-long subscription for monthly access to the rock climbing wall. Just so that I could be sure I would see her."

Alyssa smiled, shortly realizing that she was a quarter of the way up the wall at this point. The climb was getting harder for her.

"I made myself wait to tell her that I loved her. I wanted to tell her after the first date. Instead I waited three. And thankfully she felt the same way. It probably took her a solid minute to respond. It could have been a couple seconds, if we are honest, but when she said she loved me back, I knew that was it for me. She was my one and only. And she has always been that way."

Alyssa was halfway up the wall now. Sweat beading down her face. She was eager to hear more, and knew she would only get as much as she progressed on the wall. She took a breath and then another step.

Her father continued. "She was way too good for me, you know. I was a lonely foot soldier and your mother was in college and going places. I was sure she wouldn't choose me. But she did. So every single month we would meet here. I would watch her climb, help her through the wall, and then we would go eat and talk about our future. When I had enough money saved, we got married." He paused and considered the memory a moment. "I never thought I would get married. I didn't have parents growing up... Well, you know that. Had no thought of having kids of my own, and since I never had a family, I never considered one of my own for myself. Your mom was the first person I ever wanted any of that with. And I got you out of it. I think that is the closest anyone can be to winning the lottery."

Alyssa reached out to touch the top of the wall.

The phone that began to ring was so shrill that Alyssa jumped out of her chair, the memory fading away rapidly. Following the sound of it, she wandered through the kitchen and into the laundry room, no idea where she had put the thing. Finally, she found it after shuffling in the hamper. Before she looked at who was calling, she already clicked the button.

"Hello!" Alyssa answered in more of an exclamation than an inquiry.

"Oh dear..." A recognizable tone came over on the line, one that made Alyssa roll her eyes. "Did I call at a bad time?"

Admittedly, Alyssa wouldn't miss this woman. Shirley was the best mother, best wife, and best at everything that she ever did. She and Alyssa had been pregnant at the same time, and since her husband was one of the engineers for their Space Port project, they were always thrown together. But still, Alyssa was thankful for her, in a way, it wasn't all bad and she was sure she had the best intentions... Shirley always had time for play dates and homework. That helped Alyssa out a lot with Steven. Both the kids grew up together and they were the best of friends. But, for some reason, Shirley's voice grated on Alyssa. Always had, and she feared always would.

"Oh, hi Shirley. No no, you're okay. I just didn't know if I would make it to the phone on time. You okay?"

"Well," Shirly began in the most mousy, high pitch that almost made Alyssa cringe involuntarily. "Tim has been quite upset about Steven leaving. Paul and I were wondering if we could surprise him with a sleepover, just him and Steven. I mean, Paul told me I would need to ask you before I told Tim, because you might be doing something specific. I know I would be infringing on your time, but I really think it would be good for the boys!"

There was no desire for her son to go anywhere. Though at the same time, she knew that it may be good to let Steven have a night where he can be just a kid, and not the son of the Captain of the new flight to a new space port.

"I– ah—" she could only sigh in response, "Yeah, Shirley, you know what? That sounds perfect, actually. Let me ask Steven just in case. I have no idea if he'd even want to leave his room if he doesn't have to." Alyssa ended in a laugh.

Walking through the rest of the house, she went down the hall towards Steven's door, then knocking quietly. "Steven?" The door was cracked only slightly, not opening it all the way.

"Yes, mother?" Steven called back. There was the sound of shuffling as things were moved around the room. A moment later, he opened the door all the way for her and smiled.

"Tim's mom is on the phone and wants to know if you are up for an impromptu sleepover. Would you be interested?"

"Are you saying, I can?" He asked, a glimmer in his eyes.

The hope in his voice told Alyssa everything she needed to know. "Yes, Shirley, Steven would love to go. Do you want me to take him over?" As she spoke on the phone, Steven hugged his mom tightly, then moved to grab his sleepover backpack.

"No, actually, Tim and I were at the store when we concocted this plan. We are almost at the check-out line though, should be there in about ten to fifteen minutes." Shirley answered with a gleeful sound to her voice.

"That sounds perfect. I'm sure that is plenty of time for him to pack."

Just as Alyssa disconnected the line, there was an immediate frenzy of Steven saying, "I love you! I love you! I love you!" As he squeezed her tightly, arms around her waist.

"Okay, babes. I need you to get your stuff together." By then, he had already let go of her and was prancing about his room wildly. There was no attention given to her. Alyssa called out again, louder. "Okay, kid! Five second breather!"

Immediately, Steven stopped in his tracks and breathed while counting out loud.

"One." Inhale, exhale. "Two." Inhale. Exhale. "Three." Inhale. Exhale. "Four." Inhale. Exhale. "Five." Inhale. Exhale. He smiled bashfully and released the tension in his little shoulders. "Sorry, mama."

"Shall we try this again?" She glanced at his direction, and he nodded in return. Squatting onto her haunches, Alyssa looked at him, eye to eye.

Steven beamed at her, such a sweet smile that warmed her heart. She smiled right back.

"I am very happy you are going to get to go to Tim's house and have a sleepover. However, I need you to do the following things, or you will not be ready in time. I will not have Shirley wait for you because you can't control yourself. Are you ready for your list?"

With a broad grin, he nodded, giving her his full attention.

"First, I need you to pack clothes for tomorrow. This includes underwear, socks, shoes, and appropriate clothing. Second, I need you to pack your pajamas and your sleeping bag. That is it for the clothing items. Do you want to gather that portion first?" Alyssa began to list his tasks gently, slowly for him to follow.

"You can continue. I can tell you if I need a moment." The boy answered. It always surprised her when he spoke much like an adult. Part of her couldn't help but be proud of how well he was growing up.

"Okay. Three, you need your toothbrush, toothpaste, deodorant, brush, and grab your epi-pen just in case." She looked at her watch. "This should take no more than five to eight minutes and they will be here in ten, so... If you are ready..."

Steven turned and braced himself to run. He peeked back.

"Ready... Set... Go!" She exclaimed, getting up to go outside the room.

With the go ahead given, Steven raced back into his room. Alyssa could hear him gathering what he needed in a mad fury. When he was ready with his bounty, it was all thrown onto his bed, folded, and packed it all in the duffel bag his father gave him. It was smaller than the standard Army duffle bag given when you are recruited, but it was big enough for a sleepover. Once everything was packed he went into the living room and sat on the floor with his legs crossed, presenting the duffle to his mother.

"Hey, good job! Would you do a favor for me and throw out the trash since you're quick as lightning? That will give me enough time to double check your work and make sure we aren't missing anything." Alyssa added warmly, and once again he got up and dashed away.

It didn't take long for Steven to come running back, while was just finishing folding up his clothes; resituating everything, adding his sleeping bag and an extra pillow just in case.

Ding Dong!

"Mom!" Steven was exasperated, "Oh mom this is a disaster! They're here and I have to pee!" He looked about frantically– like this was the worst thing ever, which made Alyssa chuckle a little.

"Hurry and go then, but don't forget to wash your hands. I'll stall them!" She whispered with a mildly serious tone, playing along with his worries. Giving him a wink, she watched as he ran off as fast as he could go. In moments like these, he was so much like his father.

Now, to tend to the visitors. "Just a second!" Alyssa called out as got up from the couch, slowly walking towards the door. She gave it a good twenty seconds before slowly opening. "Oh, I am so sorry for the delay. I was folding laundry and had to set it down before I could come to the door." She was pleasant as she gave both Shirley and Tim a hug to further take up time. It was perfect for timing, as Steven came around the hallway at that exact moment. He shared a knowing smirk at his mother because their plan had worked out perfectly.

"Hi Mr. and Mrs. Brackett." The boy greeted them with perfect courtesy.

"Hello, Steven. How are you? You ready to go?" The man smiled and held out his hand for the duffle.

"We're going to have pizza tonight because I know that's your favorite!" Shirley stated excitedly.

Steven was already walking towards the door when Alyssa spoke up. "Hey! How about you give me some love?" Quick as

55

before, the boy turned on his heels. He ran to his mother and threw his arms around her as she knelt down.

"I love you mama." He whispered into her ear. It wasn't that he was being shy in front of others. No, he was being genuine and it melted her heart.

"I love you too, baby." She hummed quietly back.

Steven held her for just a few moments longer before letting go, making his way to the couple.

"Alright boys," Shirley, or Mrs. Brackett as Steven would call her, announced to the two before her, "Grab the gear and let's head out."

Alyssa stood in the doorway for a moment longer, watching the car leave. Nostalgia filled the air with the overwhelming amount of 'lasts' that she'd been going through. Last time doing laundry, last time Shirley comes over, last time Steven goes to a sleepover here on Earth. This was her final day in her house. Tonight was going to be all about her and James– they were going to have a stress-free, kid-free night. So she did the only thing she could think of; she ordered her husband's favorite food then took a long, well-deserved bath.

Her moment of self care wasn't too egregiously long, with food well on the way. Having gotten out and into a robe, Alyssa had been brushing her hair out when the doorbell rang. Setting the brush aside to grab her wallet, the woman went to the entrance, smelling the food before she even opened the door.

"What do I owe you?" Alyssa asked the delivery man, someone that she had seen often but couldn't ever remember his name. But she knew him well as she knew the best place to eat– they usually ordered from a restaurant called "Slims". The owner, Big Slim, and James had been friends ever since her husband commented on how great the beef was the first time they went.

"Big Slim said it was on the house. Oh– and there will be flowers coming too. You mentioned it was just the two of you tonight so he wanted to make it extra special for you." He gave her the bag of food and turned to leave.

There was a receipt stapled to the bag with a hand written note that read, *'Alyssa, good luck on your adventure. You're going to do great. Give James and the little tyke a squeeze for me. We'll miss you down here.'* Her smile grew bigger as she read the words. Tucking the note back into the bag, she went to the kitchen to start preparing pretty places to place everything onto.

The doorbell rang again, and this time upon opening the door, Alyssa was greeted by two dozen roses. She gathered them with a bashful twinkle, brought them inside and put them in a vase. She returned to the food left on the table and covered them in cloches before heading back into the bedroom.

Alyssa sat herself at the vanity and began to apply makeup, figure out a 'face' for this special occasion. Tonight, she wanted to don something that'd make her feel like a *wife* rather than a *captain*, perhaps the last she'd ever have like this. The bathrobe had been discarded for a silken nightgown, the only garment in her possession that was even the least bit sexy. It wasn't exactly revealing or intended to be seen in that way, but she enjoyed wearing it when she could, and just as she placed her foot into the last slipper, the door had opened.

"Honey? I'm home!" James called through the threshold.

"I'll be right out." Alyssa chimed, adding a bit more mascara while smiling to herself.

Naturally, James began to meander through his home. He hadn't been expecting anything different, but– "Wow!" He exclaimed, finding his way to the kitchen, "What is all this?" He poked around the living room corner to see her coming down the hall. His eyes grew big when he saw what she was wearing.

But before he could get ahead of himself, he'd have to ask, "...Steven?"

"At Shirley's for the whole night." She grinned.

"Oh. Well, then." James had already made his way to hold his wife's face, giving a kiss on her forehead. "I have no words. You are truly amazing."

"Before you drool all over yourself, are you hungry?" She couldn't help a sly smile. He made her feel so wanted.

"Yeah, actually. Let's eat first."

They walked to the kitchen hand in hand. Looking at the table and then at each other, they couldn't help but fall into each other's arms as they hugged each other for a solid minute.

"Sit down, babe. You did all this for me, the least I can do for you is pour you a glass of wine." James pulled out the chair for her to sit down.

"Well, thank you, hubby!"

In the fridge, there had been a bottle of their favorite wine chilling for quite some time, saved for special occasions. James went to pour them each a slightly full glass and came back to the table. After having placed the glasses beside their plates, he sat down and lifted the fancy cloche. On the plate was a mouth-watering beautiful steak with asparagus that was baked to perfection with feta on top, and a fully loaded baked potato.

"Oh! This is Slims– it *has* to be Slims. Oh it smells like Slims...!" The man sounded ecstatic as he grabbed utensils.

"It's most *definitely* Slims." She giggled as she lifted her own cloche with the same food prepared on it.

The night was blissful. It was the perfect evening. Alyssa thought about how every once in a while, things worked out for the best. Together they laughed, and talked and reminisced. They spoke about their upcoming adventure, and back when they started dating. When the wine was finally gone, and the chatter slowed down, James got out of his chair, pulled Alyssa out of hers, and together they walked with hands clasped towards their bedroom.

CHAPTER FIVE

Being back at the main port definitely had its perks, Jeff thought as he walked up the stairs to greet one of his very few friends.

"You smell terrible." Alex had told him in jest as they embraced in a short manly hug, clapping each other on the back twice before pulling away.

"Well, I haven't had a shower yet. If you recall, I asked if I could, and you told me no. I also might have passed out outside after drinking all the alcohol left in my house yesterday, so..." An awkward smile.

"Passed out? Outside?" The man tried to hide the amusement in his voice. "You hate the sun!" With this, the burst of laughter couldn't be contained, echoing throughout the foyer and into the *thankfully* empty corridor.

"Yeah, well, hungover is technically the term of what I still am. Last night was bad. I don't have any plans to do that again any time soon."

They began to walk through the foyer, Alex clamoring along about Jeff's ripe smell. They reached a corner and Alex tapped Jeff's shoulder in a, 'we need to get on these elevators' type of way. The silence in the building was nice. When they got on the elevator, they remained quiet too. And Jeff was thankful. It was not uncomfortable. It was just quiet. Peaceful.

Not that it really stayed that way for long. Quieter it was, the more a person could focus on their other senses, and... "I just... man, you really do stink."

Chuckling, Jeff answered with, "Man, I told you I needed a shower."

Doors clicked open. Couldn't have sooner– they were in there kinda *hotboxed*, but, it'd go without another snark. Instead, "Alright, alright, come with me," Slightly too quickly, Alex strode on as they continued down another hall. "You have to do this meet and greet, but at least go into this restroom and try to freshen up a bit."

Jeff stepped into the bathroom he was directed to and spent some time freshening up. He felt a lot better after a quick wash of the face and neck.

"Here." Alex held out a pack of gum and some Visine to the man. Jeff took the life savers, popping one into his mouth, and they began walking again. They came to a room with a door that his companion opened for them both.

Inside the room, five very serious people were standing in front of him. Formal introductions were quickly made by Alex, and then everyone sat down at a conference table.

"Hello, Jeffrey." Someone behind him spoke up.

"...Hello, Clinton." Jeff sighed, greeting a guy he didn't really like– but who was above him, unfortunately.

Clinton was known to the other people in the room. He was fair, most of the time, though, he was hard-hearted, too. His being here meant more than just a meet and greet was

62

ONE WORLD UNITED

taking place. Clinton was one of those people they called when departments needed to determine if someone ought to be fired.

"What are you doing here?" Alex said, looking at Clinton pointedly. He turned to the other people in the room. "This is supposed to be a re-admittance meet and greet. I was told this was going to be informal and light. But this is looking like an interrogation."

"Things change, unfortunately." A woman answered. She was wearing a very over-pressed pant suit. She had brown hair that was dull and cut into a bob. Her nose, Jeff noticed, was v-like and small.

"First, we are going to address the elephant in the room." Clinton stood taller.

Alex glared at the woman. He looked like he was going to stand up and march out. The problem with this scenario was that Jeff's entire life was this job– and if Clinton decided he needed to be let go, where would he go? He would be a pariah in his own field. Jeff's palms began to get sweaty.

"We all heard about what happened at the conference." The man began, slow, deliberately so. Without further elaboration– *god*, Jeff hated how he always had dramatic pauses.

"Okay? You know damned well I wasn't there– so why are we talking about this?" It was a demand; asked, even though he knew.

"Here is what I know." Clinton paused again for such a long time that Alex and Jeff managed to look at each other twice in a quizzical manner. Obviously, he wasn't stopping there, was he?

"Yes?" Alex encouraged, unwillingly.

"You screwed up. You went to see Justin. Lucky for you, no one really knows who either of you are. But, we do. Now there are

63

people asking questions about Justin. Again, luckily for you, we have an opportunity to make this turn in our favor."

Jeff raised a brow, not really seeing where this was going. He'd offer, "PR?"

"Public relations. See, because of Justin's stupidity, people now want to know our name, and what we do, and what we are all about. Our website has had a traffic increase of over 40%. Sure, we'd be celebrating if the masses were asking about our goals, what we actually *talked* about, yes– but you know they're not, it's about his *nonsense*. It's not trivial. And for us, *as we do know you,* your actions mean more than you think." Clinton paused again, raising his finger and letting it linger there. He looked at each of the five other members and nodded to the woman who spoke previously.

"Basically, you're on probation." She spoke up, putting her hands in a steeple.

What? He clenched his jaw. "Probation? For going to see my friend who *obviously* lost it over there? I–"

Faster, the official cuffed any protest with a growl– "*Look,* the fact of the matter is, we don't know how you feel about the situation, and if you are as compromised as Justin is... Well, let's just say it this way: you were off the mark by visiting, and because of this, you are going to abide by our new rules for you, and be happy about it."

"What rules?" Alex hissed through his teeth, begrudgingly.

"You're going to go see our therapist, which is on location, twice a week until we are satisfied in knowing that you will no longer lose your nerve." Clinton started.

"Are you serious?" He seemed incredulous with the answer. Jeff couldn't help defending himself, a growing anger wracking his throat. "Look, Clinton, I get you're concerned, but honestly, I just went to go see what he was thinking. I went to see if he was okay, and I went to tell him he's an *idiot*." This was so much more than a 'meet and greet'. He didn't expect to be reprimanded for just– what? Going to check in on someone he has known for his *entire* adult life?

"You don't get a choice, Jeff." Still, the man spoke pointedly, unmoved. "These conditions were only agreed upon by One World United after we told them how much we needed you. I'll have you know, they want you gone, and we compromised to keep you. *This* is that compromise."

"Why would they care that I went to visit Justin?"

"Come on, Jeff. It looks like you went there in solidarity with him, or to uh–" He waved his hand in the air, trying to figure out what word to use, though Jeff was sure he already knew what he was going to say. "–cahoots. But the truth of the matter is that it doesn't even matter. Trying to find someone with your computer skills with the clearance you already have is slim to none."

"You only have two options here." A different person with a thick Spanish accent added in. "You can either abide by our probationary rules, or you leave. What shall it be?"

They all were staring at him. Even Alex. Jeff felt their eyes on him, and felt his palms dampening where he pocketed them. It was so quiet in the room that even the slightest breath could be heard. Jeff knew he didn't have a choice.

"I will abide by probation." A dull, to-the-point answer, finally.

"Well okay then." Clinton clapped his hands together. "Now that the ugliness is dealt with, welcome back!"

Sure, his mood changed. But others were still guarded. "Are the contents of his meetings with the therapist going to be on display to you guys?" Alex interjected.

"No. The only time they are allowed to tell us anything about the sessions is if the therapist believes that he is a danger to himself, someone else, or the organization. Barring that, everything is just between them, as is standard." The woman answered.

"Also," the Spaniard added, "There is nothing that is going to be put in your file about this. We are not reprimanding you. This is a compromise to our governing body. This isn't a personnel issue."

"Mmhmm..." Alex hummed blankly. Somewhat weakly, "Welcome back, Jeff."

"Do you happen to have an update for us?" This time, it was a thin, older man wearing colors too bright for his face that inquired.

"Well," Jeff at last switched gears in his head to work mode, "The only new thing, really, is that we have the final coding for the new site specs. We are now officially a go since your approval of them. We are, thanks to recent events, canceling the public meet and greet, and having it as a closed session. Tomorrow we are double checking our go's and no go's for Wednesday."

"When will I get the diagnostics report?" Though it seemed he was asking Clinton, it was directed to the entirety of the room. Clinton was just the one whose voice would matter most at the end of the day.

Alex noted curtly, "They should be done by 1700 today."

"What is the new itinerary then?" Jeff asked, fetching his most essential items; a pocket notebook, and a pen.

"Ah... It's going to be e-mailed to you by 1700, but tomorrow the Ambassadors from One World United are meeting in the green room. So, we have blocked the entire day for that. We will be on level three lockdown soon. The meet and greet will be for the crew. We will start your double check of the codes at 0700 tomorrow morning for you and our people. Tuesday, all flight families and personnel come in for final med clearance. Then, the big dinner, uh– guess it's officially called the banquet dinner, with everyone. There will be an invite for *everyone* to be there. To say– yes, you are expected to go to that. Time is 1830 for that. Ends promptly at 2000. Lights out at 2200 for the crew and all." Alex took a breath and rubbed a hand on his face.

"Then it's go time." Jeff stated.

"Yeah, that'd be it. Wednesday, we launch. Set time for us is 0500, but actual launch is 0830. ETA on site is about 1000. We will be radio silent until 1700 when the systems will officially align with the satellites." Alex waited a moment for his friend to stop writing on his little notepad. "And then, it begins."

Jeff nodded, making sure to show no real emotion. The meeting was coming to a close. Everyone was waiting for it, shifting in their seats waiting for anything else to pop up, and not so secretly hoping nothing did.

Signaling to leave, Clinton led the ending act. "Thank you for coming, guys. Next few days will be hectic. Tension is going to be high. In case I am too busy, or not around when they land, I want to tell you now that I appreciate all of your hard work." Though he would not close on that note. No, even after 'agreement', he made a point of looking to Jeff, harsher than the words he'd speak, "Thank you for being a part of this. And... uh... I guess, just– good luck."

67

Everyone stood up in unison. Jeff did the polite thing, and shook everyone's hand.

He and Alex stayed behind, as did a couple others. This was not uncommon for them. After any 'official' meeting, there was what was known as the 'second layer' meeting. This was primarily for specific people in a specific field that talked to each other about things like small bugs in systems, or stumbling across info of note.

"We don't have time to move to a different room." Clinton informed quickly, and then moved on, figuring— *knowing* that would be enough for the lingering to begin to settle back in. He took out a small box and set it on the table. "This is a transmission disruptor, in case this room is bugged." The explanation continued as he re-opened his shut laptop. "Cameras are also cleared out of this room, so we are good. The topic, if anyone interrupts, is outfitting Jeff's new place with access to Space Port. Any questions?" He looked around. That was ominous, he felt, but it made him more curious than anything else.

No one said anything, instead leaned in.

"Okay, the actual meeting I need to have with you has officially begun." There was a beat between sentences and it was all Clinton needed to get instantly pissed off and direct it at Jeff. "Jeff, what the actual fuck, man? I mean seriously. Did you think nothing would come from seeing Justin? You'd achieve anything there? Now everyone is going to be watching you like a hawk. Did you think we needed that? I mean–"

Alex cut him off and stood up. "Okay now calm down. It's done. This isn't going to help anything. Like you said, tensions are going to be high right now."

"No offense, Alex–" Clinton began raising his voice, "–but shut the *fuck* up. I already argued for your boy. I gave the OWU a great line of bullshit to not fire him. But right now, I am not interested. See, Justin was at that meeting arguing about human rights and ethics! Brought up family safety for crying out loud! Now we're on the radar. I can pretend all I want that the media being fixated on us is a 'good thing', because of the coverage and the knowledge and the popularity base, but it was something we could have really done without. It isn't actually gonna help us *at all*, but if you put all of that aside," A finger was raised to be strictly pointed at Jeff, "Just because Justin is gone now, doesn't mean that they won't eventually connect you two together. And when they do, I am going to have to try to explain why one of our own– no, *two* of our own believes that this mission is bad." He shook his head.

"Look, I–" He wanted to continue, but Jeff cut himself off before he got very far. He didn't want to bring it up any further, but he still was *not* sorry for seeing Justin. It was the right thing to do.

"Before now, we presented this as a new resource for our planet; the discovery and exploration of more options. We were low key to see how people would take to the idea." Clinton finally noticed that he was tapping his pen against the table, and gave Jeff a look– an exhausted one.

69

Jeff finally realized that this guy really wasn't just being an ass for the sake of it. He lost sleep over this. He was trying to regain control that had been stripped from him in the one outburst that happened from Justin.

"It isn't like I am expecting an apology, though I expected it was known that when you are publicly stupid, it reflects on the whole of us. I am hoping while you had that lapse of judgment, you got some information from him that you needed, or... anything, other than just making sure your friend was 'okay' and compromising this mission. I just... I need to know you are still on board with us."

For the first time, Jeff looked around him. Really looked. The people that stayed behind for this second meeting were all looking at Clinton. He apparently had the power to sway the OWU to allow Jeff to keep his job.

Jeff had always been a quick thinker. And right now, all of his thoughts were circled on him needing to be kept in on this project. He had to save face.

"Look. I went to see Justin about the mission. But not for what you think. Before Justin cracked, he and I were talking codes. He said he was worried about a fragment. I couldn't find it, no one else could find it, but he said it was there; a strand of code that links Space Port to base that was fragile. He was working on it." There's a glance to Alex.

"Why didn't you tell me? Like that just wasn't important for me to *know?*" His gaze was returned, but the other man was less than pleased to do so right now.

"I didn't know if it was true or if he was just cracking and needed something to be wrong." Jeff sighed, looking down at his hands.

Clinton didn't so much ask, as dead-pan demanded, "So what did you guys really talk about, then?"

"I asked him about the fragment. If there was a fragment that was softer than the rest of the code, that only Justin could find. It is possible, by the way. I am great at what I do, but even geniuses can miss something small. So I asked him about a work around– how that would even work. Where I would go, code wise, to create a back door–"

"Did he answer?" Alex muttered.

"No. He said he was done talking about it ever again. So I left. That is the whole thing. I hate jails. They give me anxiety... but I needed to know. So now, I have been double checking and rechecking everything I know and everything I have done is to ensure that he really was just as cracked as he seemed to be. I'm not crazy, and my personal beliefs don't play a part here. Bottom line is, my work speaks for itself. I know what this mission is and I can ensure that my results are lasting. I am here to guaran-fucking-tee it."

All was quiet in the room. Everyone was considering, not only what was spoken, but the man that said them.

"Okay. Besides what we have already demanded for penance, I am going to try *really* fucking hard to forget this ever happened. Deal?" Clinton stood up, tipped his head to the people on the other side of the table, and moved to walk out– with one last quip, of course. "And Jeff? Take a shower."

Everyone else followed suit. The two friends took up the rear.

"...Well. I wasn't expecting that..." Alex whispered to Jeff.

CHAPTER SIX

"Hmm?" Alyssa groggily stirred under the blankets.

James spoke softly, kissing her head. "I said we need to get up soon and get ready to go." He smelled fresh from his shower, still humid from the heat radiating from his body.

"What time is it?" Her eyes were still closed. She wasn't sure she was ready to fully wake up. Time permitting, she could just roll over and get a few more minutes of sleep in. After all, she had been dreaming of her youth again and wasn't quite ready to leave those memories...

Her father had rushed into the hospital and threw open the door of the examination room that Alyssa was in.

"Where is my daughter?" Is all she heard.

He looked crazed. His eyes were bloodshot, and was shaking from head to foot. Her father was angry; so volatile it almost scared the young Alyssa– she was only eight, she had never seen him so upset. But then, she knew everyone was. The man came to her bed and scooped her up. He held her, and for the first time ever in her life, Alyssa felt her father tremble. He choked a hard breath, one that spiraled into a breaking sob. She knew she had been in shock, but she remembered feeling like she... should have done more.

Alyssa had been heading home with her mother. After having a horrible day at school, her mom had taken her to see a movie, grab some pizza and then ice cream– average reassurance for a girl her age. But things couldn't turn out so ideal, never so straight-forward.

They had been hit head on by a drunk driver when on the road.

By some miracle, Alyssa, having sat in the back seat of their car, only had a mild concussion. Her mother... Oh, she was not so fortunate. Crashing like this as the driver... No, she could not be helped. The woman was killed on impact.

In ways she couldn't even fully comprehend yet, Alyssa's world was forever changed. Her body ached and she was confused. No one was talking about her mother, even if she tried to ask about her, beg weakly for an answer. Through those hours, nothing was said. The girl hadn't seen what happened to her.

That was the way it was for much of the rest of her life.

They weren't the kind of memories she liked. Ah, these, they stung. The dream faded away as her husband began to speak, Alyssa properly waking. "It's almost nine. We're supposed to pick up Steven at ten-thirty." James sat on the bed, bracing an arm on either side of her body.

She stirred a bit quicker at the mention of her son's name. The woman rolled out of bed and kissed James' cheek before heading to the shower. She could hear him moving around and knew he was getting dressed before getting some coffee. That was routine, anyway. He'd always been a morning person. The military made her a morning person, too, but she wasn't like that by preference.

Still, James probably already made all the last minute preparations for leaving. There was no real reason to come back to the house once they picked up Steven, so they were making arrangements to have everything dealt with directly after they left to go pick him up.

"Toothbrushes?" Alyssa asked after her shower, coffee in hand. The nerves were starting to build in her.

"I grabbed all of our bathroom stuff while you were getting dressed. And I checked Steven's bathroom. That kid took everything except the toilet paper." James giggled.

She nodded. An uneasy hand rubbed the back of her neck, but, *it was fine*. It was all going to work out fine. Still, Alyssa blurted— "I love you."

"I love you too, babe."

"Uhm, did you check his room? I want to make sure we take all the important stuff. Maybe we should make a list..." She trailed off.

"Babe." James said, grabbing her by the shoulders to help her stop reeling. He was great at that. "It's okay. Breathe. Look at me."

It wasn't like she was having a panic attack. She was just worried. The same thing happened when they were moving to base. Fretted over everything that was packed, where it was packed, and what they needed for essentials until they could unpack. This was different, though. It wasn't like they could go to a CVS if they needed something— but also, the base would have stocked anything they could possibly need for the next decade or so.

Her husband knew how to calm her down regardless. He knew how to reach her. Alyssa was great at the military stuff, but she

75

was nervous about everything mundane, that part of her life as a wife, as a mother. Tears welled in her eyes when she looked up at the man.

"Hey–" James pulled her into a hug. He was tender in his each motion, careful as not to smother her, especially as he held her closest. "What's wrong? Talk to me."

"What if he was right?" She whispered, a tear rolling off her cheek and onto James' shoulder.

"Who was right?"

"The crazy guy at One World United."

"The... crazy guy?" He almost asked *who*, but... While confused at first, comprehension started to form.

She couldn't explain it. There was so much emotion and so much at stake. Her entire family, in fact. Everything was perfectly balanced in her world until that meeting. Now she felt like there was a shadow growing each day that the launch was nearer. She should have told James about the nightmares. She never did move much in her nightmares – no indication that she was experiencing them – so he never knew unless she told him. This time, however, she couldn't put it in words. Could this just be stress mixed in with the uncertainty?

"Are you really worried about what he said?" James asked calmly, gently as possible.

She took a deep breath. And then another.

"It just threw a wrench in my 'perfectly balanced world', and now I am wondering if I am putting my faith in the wrong people. But I know that's crazy. I just... I know how to do this on Earth. Though, up there? If something critical happens, that's it for us, isn't it? And that's scary enough without thinking that there's some secret plot they have that'll rip our family apart."

"I will say... I can't help how you feel. You are entitled to your emotions," James began, rubbing circles on her back, "But I can tell you that whatever the future brings, we will handle it together. And I can't promise everything will be perfect, or go the way we want it to. I wish I had a magic ball, but I don't, and we don't believe in that stuff anyway. I know how you fly. And I know how you captain, too. I am *confident* that we are going to be okay. I have faith in you– your capabilities and your person."

"The most dangerous part of the mission will be us getting there. Besides that, the only other concern are day to day struggles. That's doable, that's planned for. And, I can handle the launch. I *will* get us there." Steadily, Alyssa nodded, "...you're right. We can't know what the future will bring. We can only know that we do the best we can."

"Like we do."

"Like we do." She answered.

They got Steven right on time, and then headed towards the center. The ride was quicker than she anticipated. But her head was back on track. She could handle the task in front of her. The guy at the conference was just that– *crazy*. That was why he was locked up. She was aligned with her universe again. She was in control.

They were going to have breakfast on base. The place was popping, with people milling about everywhere.

Their son sounded out from the back seat, "Mom, do you know if there will be wi-fi? I need to follow the media coverage for our trip. I also want to look up a bunch of random places and put the pictures on my tablet!"

"Why are you so concerned with our media coverage?" Alyssa asked while looking around the parking area they were entering. They were going to be getting a room in the center for the night and then tomorrow they'd travel out.

Something, something about looking super smart on TV, and then more about– well, *a lot*. Steven was talking nonstop all the way to their room.

"If they ask me questions, I want to know how to better answer!"

"Answer better." She corrected, opening the door. The room looked like it belonged in a hotel. She walked directly to the fridge, grabbed a bottle of water, and cracked it open.

"That is what I meant. To answer better! I just want to know what it looks like when I talk and how to talk better."

"Okay, well that isn't a problem, we can figure that out. We won't get our passcodes until after breakfast, though. Our clearance isn't active until then. Hm, but right after we eat I will grab them, okay?" Alyssa answered.

"Okay, mom. Mmm, I hope there's bacon." He said. Enthusiastic, yes, but something else seemed to possess him, something Steven thought about for a minute. "Hey mom?"

"Yeah?" Alyssa answered.

"Can I have some coffee?" A spark of hope– or *fire*, whichever it may be.

"Not today. You don't like coffee anyway." It was doused with a half-hearted answer.

Regardless, he tried, "Yeah, but I'm going into *space* now. I am practically an adult. I mean, I'm leaving the world behind. Literally. That should allow me to get one cup of coffee!"

Of course, he wasn't getting what he wanted. Though... it stirred something else. "Oh baby. Do me a favor? Don't grow up so fast, okay...?" It drew a sad smile she tried to hide. Steven wouldn't notice, no—his father stepped out of the bathroom with more than appreciable timing.

"Dad!" The boy called out to his father as he mulled into the room. "Mom won't give me any coffee. Can I have a drink of yours?"

"Ha! I haven't even had a cup yet!" James tousled his son's hair. He, too, was disregarding their son's call to caffeine. In lieu, he turned to ask his wife a question. "What's on the itinerary today?"

They were back out of the apartment and heading towards the elevator. It took a minute to find the mess hall, but once they did, they were not sorry to see it. There was an entire coffee bar, followed by rows and rows of buffet style breakfast foods in silver rounds. They eagerly loaded up their plates and found a table to sit down at.

"Oh!" A girl piped up from a distance. "You're the captain!"

"We were just talking about you! I'm David. I spoke at the One World United conference!" The man beside her had introduced himself.

"I am Jessica!" She stated with shared enthusiasm.

"Well, hello. I would be happy to answer any questions and talk to you, if you would like. But, I would prefer the mindless conversation that usually comes with breakfast." Alyssa answered the two politely.

Jessica replied cheerfully, almost too quickly, "Actually, we are pleased with the pointlessness– mostly. We were just talking about the coolness of actually colonizing a piece of space. I mean, ten years of a trial is long, but people stay in the exact same job for twenty-five years, and usually those jobs are super boring. This is a super important military mission that you get to take your family on! That's super cool."

"I actually was thinking something similar this morning. There is a very real danger that comes with any mission, but I'm lucky enough to have my family at my back." Alyssa responded, taking a piece of bacon and eating it. James smiled and rubbed Alyssa's back while drinking some coffee.

David concurred, "That is true."

"So cool." Jessica echoed.

"What do you think?" The other man couldn't help but ask James.

"It will all be worth it in the end. I mean, how can I not follow the woman of my dreams? Anywhere she goes, I do too. For her, I don't think the edge of the Earth's atmosphere, or even beyond, is all that far. She has worked so hard for this. And while we are up there, the only thing that will be different is that we will be seeing a lot more stars than you will."

"Well how cute is that answer?" Jessica hummed, batting her eyelashes. "How do you feel about it, Steven?"

Everyone looked at Steven who was eating his cereal with a look of bliss. They waited for him to chew, and it seemed he was too, trying his best to be polite about it all. Just as soon as he could gulp the bite down, though, he showed his teeth in the widest grin.

"I'm too young to have too many thoughts on it. I can be okay with anything! I'm resolent."

"Resilient." Alyssa corrected.

Steven practically *hoorahed*– "But I am gonna be in space. I'm excited! I'm going to space!"

The whole table laughed and, true to their word, they spoke of nothing but random things for the rest of the time.

Alyssa needed a moment to herself after that. The day was a blur. She met so many people and was ushered from one room to another. It was impossible to keep up with the names and the faces. Finally, they were going to the control room for photos and then back into the cafeteria for lunch. It was barely time for that, but Steven was excited and nothing seemed to bother him, and James was an absolute gem. She hoped, silently, that the day would start to go by faster– once lunch was finished, they could finally take a breather.

CHAPTER SEVEN

L ongest. Night. Ever.

Jeff couldn't remember a time when he was literally so drained of energy he didn't even think sleep could fix it. In fact, he didn't sleep– though it wasn't for lack of trying. Instead, he just tossed and turned and stared out of the window of his new house. He was unable to find the bedroom things with all of his unmarked boxes. Forced to sleep in his boxers, and to use a jacket for a blanket. No pillows, either– and it hadn't been the night for no pillows.

He watched the moon and the stars pass his window– watched the sky slowly lighten to sunrise. Gave him too much time to think about the past...

"Dad—just *listen* to what I'm *saying*. I could easily get into the program, and you and mom would be saved from the war and have enough money to start fresh, and do anything you want!" Like it was yesterday, he was yelling at his father again. Some desperate fit to fix things– cliché, huh?

The war had been raging for some time now. People were dying left and right. They were the lucky few unscathed– thankfully, Jeff had already finished MIT and was working for an independent firm that dealt with codes for planes. Somewhere along the way, he was approached by some 'new regime' who were looking for the brightest and the most helpful people, collecting them all for the new world that they were creating. It... sounded good. Like a chance to make a difference kind of good. What they might've *needed* kind of good. And delusional, too; he wasn't stupid.

Anticipatedly, Jeff's father hadn't wanted to be a part of whatever regime this was. He was an older republican in Wyoming, and Jeff knew his exact thoughts on the matter. America! He would shout. As if that answered all.

That afternoon, he never got a straight answer from him. Never would. The war had spread too close to his house— an air strike had left his entire city in ruins. They never found his family, and he'd had to face the reality that he was to make choices on his own. He'd been devastated. So when the new regime came to him again, he followed that 'too good to be true' with hellfire in his eyes— he decided to do it. *Anything* to end the god-damned war sooner than later. He didn't even care who'd win.

Jeff had met Justin at the new center he was working at. They went into the program together, and the only one that he could admit was equal to him in codes had been that man. And no, he was not being pretentious. Codes made sense to him the way a piano made sense to Mozart. They laid down for him like the lovers he never had; he could manipulate them any way he wanted. It was his perfected art, he was the team's virtuoso.

Though, despite it all... Justin was their anomaly, too. He said he hated code. But, they were easy for him. What stumped Jeff sung to Justin— how could he *hate* this? He was the go-to guy for the best of the best, and that included himself.

Eventually, they found themselves working on everything together.

By the time the war had finished, Justin looked entirely different than the man Jeff had originally met. Exhausted, with dark circles under his eyes and weight slipping off so fast he was becoming a skeleton. Jeff was glad the fighting had finally ended– relieved, even.

But, where there should have been celebration, Justin... was on edge. Kept saying 'it' wasn't okay– whatever 'it' was.

Whatever joy he must've felt was short-lived, and little expressed. Despite the global ceasefire, Justin only threw himself deeper into his work, as if it wasn't already eating him alive. Doubled down on his efforts and into programming for the space station. With every passing day– *hour*- Justin's decline became more noticeable, but Jeff couldn't do anything to help his friend. He could only watch as he kept losing sleep and truly wore himself to bone. There were still many times Jeff tried to ask what he was doing, or, *how* he was doing. No work talk, no, not anymore. The only response he'd ever get was that it wasn't okay. It was not okay. *It was not okay.*

Over, and over, and over again, the only conviction left in that brilliant mind.

Jeff's alarm went off abruptly.

"Well, here we go..." he muttered to himself as he braved his way up from the bare mattress.

He was headed to shower, as was his normal routine, but stopped mid-step. Foot hanging in the air like a flamingo, he thought a minute about what he would need for a good bathing. Then, he switched directions. His target? The kitchen. He needed a key. Maybe... his truck keys. Or a knife. Something to open up as many boxes as he could. It was easy enough work to open a box, see what was inside and move on. To actually get a good rinse, he needed supplies. Toothbrush, shampoo, conditioner, razor, a towel, and... then he could find some clothes. That'd be awesome.

Jeff was cursing to himself. Of course, it is usually the way of the universe, that when you have a bunch of boxes in front of you, what you are actually looking for and needing will be the very last thing you find. Thankfully, he still had some good time before he had to leave, almost as if he expected this turn of events... He couldn't be late today.

Now that he had everything he needed, he stepped outside for a quick drag, looking at the sunrise and beginning to feel a little normal. He guessed he did sleep there at the end. He was dreaming of his father. How this all started.

"Oh, Pops... I miss you." He murmured, crushing the butt of his cig on the bottom of his shoe before heading inside. He threw it in the trash and went to shower.

That shower was short and sweet. Decidedly fitting, he wore his lucky gray pants and a teal shirt with black shoes. Everything was... semi-wrinkled, but his mother would say, *"There is naught to be done about that right now, my son."* He went to the boxes and searched some more. This time, he knew where he was looking and what for. He needed his spare gum and breath mints. As weird as it was, those were as on him as a pack ever was. Which, speaking of, he threw a new pack of smokes in his jacket pocket; slipped the leather over his shoulders and tied his shoes again. They had been too tight the first time.

This was his life's work. This was his everything. He was completely confident that he could handle today.

People didn't know who he was, or even what he did, but his work still spoke for itself. And there was a difference in having confidence and arrogance– he knew things could go wrong. But he was also confident he could fix them.

Jeff got onto the Space programs radar by doing side work for NASA. It had allowed him to pay for college, since MIT was expensive, and his parents were just about the average American middle class people, nobody special or rich. NASA loved Jeff's ability to dissect and explain the same topic from a multitude of approaches, depending on their client's– or really, any person's– understanding of the subject.

That especially was why he was excellent with reports. He could have the same report for three types of people and they all knew what he was saying. It was because of NASA that he learned some other tricks too. To say, being around all sorts of smart people helped him to learn how to not fidget during meetings... taught him how to keep his head high and keep his cool when he was being reprimanded, and likewise, how to look someone in the eye and tell them to fuck off, anxiety be damned.

He pondered all of this while multi-tasking his many-layered lists of tasks on 'to do' for the day– the next few, even. The man headed out to his truck, started driving to work, and went on thinking about the math. He was mostly confident about everything because said mathematics never lied. They couldn't, it made sense; chuckled to himself at something Justin used to say, "After all, the math checks out."

"Feeling alright, Sleeping Beauty?" Alex asked with an eyebrow raised.

"Actually, yeah. I'm starving though. And I need..." Jeff was looking around trying to find what he was looking for.

"Coffee bar is over there. Get as much as you want– I need you calm and cool today, alright? Calm and cool."

"You know, you don't really make me feel exactly *calm and cool*' when you say it like that. But uh, grab me a bagel and cream cheese while I look at the coffee."

"Yes, dear," Alex rolled his eyes, "Toasted?" He'd still look over his shoulder to ask.

"You know how I like it." Jeff replied teasingly, not even bothering to look his way.

By the time Jeff got back from getting his coffee, Alex already had his bagels toasted to that exact golden brown he preferred and was spreading them with his topping of choice. Kinda routine, so he really did know how it was done. They both knew cream cheese went perfectly with bagels; bagels went perfectly with coffee. And although Jeff hated eating in a glorified mess hall, he loved a 'shitty' little breakfast like this, it still an undeniable merit. He could pretend he was by himself here. Just him, his coffee, and his bagel.

Alex left not long after Jeff had eaten his breakfast, though Jeff stayed for another cup of coffee. After filling it to the brim, he wandered out of the mess hall and into the area towards the elevators he needed. While it wasn't really surprising, there were people milling about everywhere. Felt very crowded... expected. But as Jeff entered his department, it finally caught up to him. There was this strange energy in the room. At a glance, it might've been because it was so *huge*. This location was second in size only to the spaceport center's main control room– which nobody was allowed in, at least not yet. They had to finish this check in first.

Time seemed to slow for Jeff as he took in the sight around him. Everyone was talking, eating, shifting paperwork and working off nervous energy. People were scuttling to their desks. The talk, Jeff was pleased to hear, was all about the tasks at hand, or those to-do. Perhaps the great perk of working in a place full of introverts, focused people without much investment in gossip, or whatever it'd be called here. Nothing was said about Justin and One World United; nothing about anything other than codes, programs and application processes to be sorted through.

All said, whatever comfort he took was soon to change: the air of the room immediately shifted when the program director and the main supervisor walked in. There were plenty of notable titles that followed. All eyes followed the group headed towards the back, where a food and drink stand had been laid out for the workers to have for today.

This entry of past Presidents, Managers, Generals, and such higher powers officially started the day.

The higher ups gathered their morning niblets and ascended a set of stairs that the crew called, 'The God Box'. All the big decisions that were handled on the ground were made in the God Box– the people occupying that room had ultimate control.

Precise to plan: when the radio lines opened up, the first checks of the day began.

"Fuel control."

"Go."

"Medical."

"Go."

The long list of marks lasted nearly twelve minutes until Jeff heard what he was waiting on.

"All pre-emptive crew codes are a go." A loud announcer bellowed into the overhead speakers.

Every person in that room erupted in cheers. They were done with their part, for now. Most of the group had nothing to do until the passengers and crew started to wind their way to the flight deck. So they'd find another way to fill the time as excitement settled down in the room. Jeff himself took a deep breath and started opening emails. He was anticipating a notice to re-work the math.

But... for the time being, all was ready for launch.

CHAPTER EIGHT

B est. Morning. Ever.

Alyssa woke up ten minutes before her alarm and with a smile on her face. For some reason, she always felt like she was a little more prepared when she woke up earlier than it could ring– felt like she was beating the day.

Now that the time was here, she was practically invincible. Face to face with the challenge, the captain had the rare ability to stare it down in utmost confidence until after it was done. Oh sure, when she landed and was able to go to her quarters, Alyssa would throw up and shake for a solid minute, letting the anxiety roll through her, and the night before she'd let herself dread terribly. But, in the heat of the moment, that time when she would be needed most, she was calm, cool and collected. See, in her mind, it was that point of do or die. There was no reason to let it get the best of her. Yes, the nerves were there in her body, just withheld now until after the fact. It was one of the reasons why she made a great pilot. She wasn't fettered until *after* she was on the ground– never in flight would anyone see her shoulders agonizingly tense, her palms grossly clammy... this was her life's purpose. She'd be damned to run from it.

Alyssa didn't turn off the alarm when it began to ring. She left it for James, heading to the bathroom for her morning routine. The thing shouted, *'BEEP BEEP BEEP'*– unprotected behind a half-closed door, she could hear it blaring as she was brushing her teeth. She'd need to withhold a small giggle as, through the open crack, she could see a very grumbly husband rolling around their bed to silence it.

"Morning babe," James yawned. 'Battle' was decidedly over as he began moseying out of bed.

"Hiya! Did you sleep well? Are you ready for the day?" She began, with a bit more enthusiasm than she'd intended.

"Shhh... A little too much energy this morning. I need coffee before I can handle all that. Besides..." Another yawn, then a half-joking huff, more fond than anything, "You're not a morning person. Don't act like it." He kissed her head and joined her in those small routines.

Chuckling to herself, Alyssa strode down the hallway to wake Steven, even starting her 'morning song' for him.

She always made sure to chirrup that familiar wakeup call. It was the one thing she clearly remembered of her mother. That, and the 'I love you' song. Sometimes hearing its chime was hard on the heart... though she felt good today. For the first time in a long time, she had dreamt of her father and woke up happy.

It was the day of her graduation.

She was finally finishing high school, and her father was completely changed from what she knew so well. His demeanor was so different– smiling and laughing openly, doing all the good things that fathers were supposed to do this day. He ironed her gown, bought her a new dress, and treated her to a spa day.

The graduation was perfect. Her father had flowers for her and cried honest tears of joy. Everything was perfect.

"I said good morning, good morning, good morning to the sun~!" The woman called out, her thoughts back in the present.

"Mmm..." Was all the reply she got from the lump that lay beneath the sheets.

"Come on, buddy. Time to get up and get ready!" She hummed, walking over to the bed and sitting on its edge.

"Ugh, mom–!" He grumbled and rolled over. "I was kind of having the best sleep of my life."

"Aye son, I've heard that before. But you know what? We are going into *space* today. You have to be up soon so we can get ready."

"They can wait for us." He muttered lazily and closed his eyes again.

She walked out of the room and came back with a cup of water... and poured it on his head.

Things moved quickly after that. They had their breakfast, and headed down to the dock. It'd been a comfortable venture for the morning, all rude awakenings aside; Steven hadn't taken it deeply personally, albeit, he was still sopping around. Regardless, it was important they do things timely and be prepared for the day, without too much stress.

In fact, they'd arrived just in time for a loudspeaker to call– "Passengers to the entrance bay."

It was a moment of stillness in their life. Here, now, had come that determining second, the one she would look the beast in the eyes... there was no turning around after this point. And decidedly, they were not afraid, they would not walk away from this. When he took the first step forth, uninhibited, Alyssa knew that.

And recognizing this, he offered a gentle smile: "Good luck, honey." He gave her a hug and kissed her, holding it for a long second.

"Be careful driving, mom." Steven added, hugging his mother as well.

"Thank you for being awesome and coming on this adventure with me. Both of you... but especially you, Steven, you're doing amazing. I'll see you up there, okay?" She kissed his forehead.

And then she watched her husband and son walk through the doors.

The world was silent for a moment to Alyssa. There was her everything, she thought, walking through those doors. The duo looked back and waved big, one more time. Beaming at her, her cheerleaders.

"See you guys soon." Alyssa said with another smile and a wave, trying to somehow be bigger than theirs'– matched enthusiasm, *hope*.

"I love you, honey."

"I love you, mom."

"I love you both very much."

They all turned in opposite directions and walked to their destinations.

Sooner than Alyssa processed what had happened, she was in the cockpit, and the countdown began.

"10... 9... 8..."

The entire vessel began to shake as the engines roared to life.

"7... 6... 5..."

It was like riding a coaster trembling along its skyward path, before the swift downward rush; anticipating the click when the railing would give way, that moment everyone held their breath for.

"4... 3... 2... 1... Prepare for liftoff."

The acceleration had even the most experienced crew member braced back in their seat. The G's that were being forced on everyone made it hard for James to even turn his head. He managed to turn left, though, and look at his son. Steven's eyes were screwed shut and he looked terrified. This was nothing like the simulations, no matter how many times they'd run it, this was so much more than any average person expected.

"Steven! Hey! Steven!" James yelled to his best ability, the pressure screwing up his jaw all throughout.

Steven fought to open his eyes and turn to his dad. He was trying so hard to be brave. This felt awful.

While he was trying to say something reassuring, anything more, that weighted jaw slacked, and "*Ag-uh-uh-uh-uh,*" poured out helplessly the very second James let his face shake to the beat of the 'earthquake'.

It was silly in a sort of jarring way. Steven laughed, but it made his teeth clatter, that *hurt*, so he opted to open his mouth and did the same thing his father was doing. And all of the sudden, it was pretty funny to be up here with their faces all smooshed together and shaking through a cacophony of non-sense sounds.

Even still, with some grasp at composure, if it could be called that, he was trying to laugh– it was so hard to do anything but that 'gargle-speak'. But, it was a saving grace somehow. The tension they'd spent so much time building up and now faced was broken. James sat back in his seat and smiled, marveling at how resilient kids really were.

This shared 'joke' was soon to end as, once the ship broke through the atmosphere, the ride became smoother.

The plan, as James understood it, was to go up and out of the Earth's atmosphere, then coast on towards the space port. He thought it would take about four hours– but then again, it could be fifteen minutes. His brain was still scrambled and askew from the turbulence. But now that he had a moment, he gave all his positive vibes to his wife. Man, he was proud of her. Having seen all the work she did to get to this point just made him realize what a total badass she was. Yes, things were a team effort in their family, but this? This was all her; they adjusted to her schedule. Because she was *the one*– the one to do it all.

The best thing he could do for her was support her.

Musings like these, everything he was thankful for, and everything he had in life sapped him out. It was helpful though. It took over the time of the trip. The calm took over every time when he thought of how lucky he was, and he hoped that if Steven looked over, he would see it and feel soothed, too. Though it seemed Steven was currently looking out the small window down to Earth. Sometimes all you needed was a– good- mind– *ough* — set.

The jerks were intense as they exited the pull of the Earth's gravity, halting his thoughts at each and every emerging word. They were almost to their new home.

This was it. This is exactly what Alyssa had been waiting on. All of her life, especially her adult life, she had waited for this— *worked* for it. And hand in hand, she had a wonderful family that helped her stay grounded, even when she was up in the sky. They'd made the journey more bearable... something Alyssa didn't need to feel so daunted by, however audacious this entire mission appeared to the everyman.

Alyssa just wasn't that, nor was her family. She was the captain, and they were honorary crew.

Her eyes saw the dials and screens clearly. They showed her everything she needed to know: plans going accordingly. Every calculation made by every person working years to set this in the making showed itself in bold, as had every bit of training placed in her hands. She couldn't wait to arrive at port, have all passengers guaranteed safe, *mission success*; be at the end of one of humanity's most perilous journeys, and begin another. But for now, it was another four minutes prior to coast time. Her focus stayed exactly where it needed to be. She within her profession was much the way you'd imagine a doctor: give orders for one thing, read the screen for something different, assess and accommodate for new information. And there were no alarms, no bells, nothing to cease function; rehearsed calm. She watched the numbers change by the second and kept analysis ongoing to coasting and beyond.

"Stabilization thrusters on my mark." Alyssa said after a moment. "Three... two... mark."

The change was immediate. The vessel leveled out horizontally and there was a moment of weightlessness that made everyone's stomach lurch. The sensation was not unlike the second an elevator started moving– a fast one.

"Forward thrusters are a go, Captain." Her second-in-command answered.

"Forward thrusters on three... two... mark."

Everything synced to perfection, just as she knew it would. There was no lag, delay, or grind as she felt the shift. Gliding into the space port allowed them to be back online with headquarters for a moment before they'd lose connection to the satellites once again. This was critical, however brief it was.

"Earth Point to Captain Burns." The screen cleared to show the room she had left to go to the docking bay.

"Captain Alyssa Burns to Space port. I read you loud and clear." She smiled at the monitor, "Clear atmospheric conditions and green lights all around to continue on to space port."

"Earth Point direct, Captain. Great job."

A nod, then confident reply: "Captain Burns signing out." *Connection lost.* Talk about perfect timing.

After the green light turned off and the comms shut down, the entire deck burst into cheers and applause. She imagined that the room at Earth Point was louder. They'd survived the hardest– the landing was the easiest part of the trip. Alyssa grabbed her first mate into a short embrace and the air was electric with excitement.

"Alright, alright, we got work to do," Alyssa passed the notion in an attempt to calm the crowd.

It did work. They could celebrate more later– for now, her crew went back to work mode almost instantaneously. People were still talking, though it was task-related and needed. Adrenaline slowly dwindled in the voices, even if a few carried on with fervor, a genuine joy for the mission, for the lives ahead of them all... And Alyssa thought that the whole thing actually started to feel real now. This was beyond her practice, the most critically emphasized pieces were over and she could almost be human again. So she'd come upon the act of the mission she was least worried about. She was relaxed, nothing was about to go wrong, just... not totally. Her old habits crept up on her again, back-of-the-mind needless worries or doubts or something or other, just, anxiety at the fringes, festering.

Sinking in its teeth: life was about to change. It'd be better, they all hoped. This was better, wasn't it?

The minutes ticked on, and soon they were docking at the space port.

"Home sweet home...!" James opened the door to a room down a nondescript hall.

The walls were not so much white, but gray. It looked like it was attempting to mock blue, just couldn't exactly be called that. Though, despite these barren and gray walls, the inside of the space was actually quite lovely. Shockingly in contrast, actually. James had been in man-camps and barracks and this was *far* superior. The rooms were spacious and colored warm, lovingly decorated with hanging art, splashes of paintings, and furnishings to match each room. The kitchen was a galley, one of a muted, burnt-orange color with mahogany cabinets. The appliances were white in the way they often were, but the contrast in this case made it all come together very elegantly.

All of their belongings were being loaded into the space. They were being situated by location based on their titles on the box. And wasn't that going to make unpacking a breeze? There really wasn't much to unpack, anyway. It was already furnished. Really it was just their personal belongings. It was kind of like an apartment, the way it was set up. But still, it was theirs. Their own strip of heaven.

"Dad, can I go wander around and get lost?" Steven asked.

"Sure! Absolutely! That's a great idea! Go make a friend." James messed up Steven's hair and turned to the first box he came to.

Steven started to walk down the hallway, then walking back a few times until he could remember how to find his home again. There was an immediate difference between Earth and the Space Port. Everything was more flat and institutionalized and also bigger, grander, and just... 'more than'. Steven hadn't thought that there would be much difference in looking outside the window and seeing the sky, or seeing space, but it was.

He also couldn't help but notice the difference between the people. Secretly he was worried that people here would only want to be his friend because his mom was the Captain. To him, she was just his mom. She wasn't Captain when she came home. She played gin rummy and liked doing puzzles. And his dad, well, he may be a beast in the gym and everyone's personal trainer, but at home he played baseball with Steven, helped him build things. He assumed most families did the same, right?

Meandering quite aimlessly, looking at nothing in particular, he let his feet take him wherever they thought he needed to go. The anxiety was growing in his head now that he was out of his own comfort zone and away from all of his friends he'd known back on Earth.

He came to a bucket on the floor and looked at it. It was weird to find the thing on the ground in a random section in the ship, wasn't it? But then again, Steven mused, it is space. Why wouldn't there be a random bucket on the ground waiting for him to find it? Looking around to make sure this bucket didn't belong to someone that was directly in the area, he flipped it over and sat down.

After he let his mind go dark, he became peaceful. He stared out at everything and nothing in particular. Finally the area around him began to take form. He was no longer in the residential section of the ship, but the more open part where real life would be lived. There was a corner that would be perfect for baseball, there was a playground, and areas that had him feeling like there was a sense of normalcy.

"Excuse me," A gruff, accented voice called from behind Steven.

"Wha–?" Steven yelped and fell over off his 'seat'.

"Oh, hey now! I'm sorry. I was working on getting my things to the right spot. My area switched and I just came back for my bucket, that's all." The man had his hands raised in a clear gesture that he meant no harm. He was a burly man. Darker skin tone, an accent that Steven couldn't quite place and a handle-bar mustache. It looked as though he was trying to appear smaller. This put Steven at ease.

Steven answered as he got up and tipped his no-longer chair right-side up, "No, I'm sorry I used your equipment for a stool. I just needed to get my thoughts in line."

"Well, thank you sir," The man managed a smile.

"—can I be nosy?"

"Ummm, about what?"

"Whatcha working on?" Steven cocked his head to the side.

"Oh! I work in maintenance. I was just doing prep for some janitorial work later. It's easier if I have it all situated before-hand." The man leaned on the wall, staring at the area in front of him.

He nodded, taking that much in. Then, too formal-like for his age, the boy outstretched his hand for a shake, "My name is Steven. My mom's the Captain. So, I sure do appreciate the maintenance you do!"

"Oh! That is kind of you, Steven. I really should get back to my duties, though," The man shifted and began to gather his bucket.

"No, please— don't go yet. Hold on just for one moment, if you can?" Steven looked so hopeful that the man couldn't help but listen. Even freeze, the identity of this child in consideration... but he didn't seem upset or unpleasant.

"I'm not sure that would be entirely appropriate, but uh, just a moment should be fine."

"Oh I was just curious about the layout is all," Steven said, looking out and gesturing to the space around them.

"What about the layout? How can I help?" The man leaned back against the wall and, again, tried to look smaller.

An odd inquiry followed: "I know there has to be better places to see than this, right?"

The maintenance man grimaced— just a flash of one, like a wince, or like he wasn't so sure of what to say, though he carried on anyway as if he hadn't had any reservations. "Well, I know of a few possibilities, but I have yet to go anywhere except from my one room to the one I am moving into. I can keep an eye out."

ONE WORLD UNITED

"I was hoping for a better place. Somewhere I can just use to think."

"Well..." The man echoed, and then laughed softly, "When I do my rounds tonight, I will keep my eyes open for such a spot. If we meet again, young sir, maybe I will have some spots for you."

"Thanks! I will find you tomorrow maybe? About this time?" It was difficult to hide the excitement on Steven's face.

"That will work. I will be around this area then."

"I should probably go back home and unpack my room. I don't want my mom or dad doing it— they never put things in the right place." He smiled bashfully, giving the man a wave.

"Have a good day, little sir."

Steven ran off back to his living quarters. He was so excited. He made a friend! As soon as he got home and saw his dad he told him all about the new guy he met and how he was in maintenance and was going to help him find his own thinking spot.

"Well, I would love to meet him!" James said with a smile.

"Sure! Can it just be our thing though?"

"Why?"

He seemed tentative about saying it. Slowly, "Mom knows everything about here, but it's fun to learn some of the secrets on our own. You know what I mean?"

"Yeah, sure. That makes sense to me. Just know if she asks me straight out, I will tell her. I will never lie to your mother... But, ah well, I don't see why she *would* ask." James smirked. "We can keep this as our little secret."

The next day couldn't come soon enough. Steven was so excited he had trouble even pretending to fall asleep. To make a friend on his very first day made him so happy. His head was filled with thoughts of helping him with maintenance tasks and learning the ship better than his mom. He then thought more of being able to find the best hiding spots and of his new friend, who could fix anything.

When he did finally sleep, morning was almost on the rise. It was a strange sort of feeling to have not realized that you slept at all, but knowing you had to have done so, because you woke up. Mom was already at work and he wondered if his dad would be busy and wide awake too. Steven felt a little guilty. He had wanted to wake up early and wish his mom good luck on her first day. He got up and had less of a bounce in his step than a lurch. His morning bathroom routine was quick and then he walked to the galley kitchen area.

There was breakfast half eaten when Steven walked in. The smell in the air was pure bacon and eggs. It was Steven's favorite breakfast smell. This was just like Earth.

"Morning, Mr. Lazy-pants." James said in a half smile.

"Grumble grumble." The boy jested.

"Uh-huh."

"I was so excited to sleep for the new day and adventure that I didn't sleep well. So when I did sleep, it was longer than I thought it would be..." The boy grabbed a piece of bacon and sat at the table.

"Well, bacon and eggs are the best cure for that, don't you think?"

The meal was so satisfying and his dad was right. It *was* the best cure. It gave the energy required to, well, have energy at all. Something about breakfast being the most important meal of the day resounded with it.

"Hey bud. I have some grown up things to do today, like find the gym and start scheduling clients," James began. "Do you want to come with me?"

"Nah. Not today, at least– you know I'll end up seeing it more times than I could *ever* possibly want."

There was a laugh that came from his father at that answer. When he saw his son begin to head towards the room, James called out, "Now hold on!" There was still more to say, of course. He waited for Steven to return, at least close enough to hear him, and pay attention. "A couple of things before you start gallivanting around. One: no finding Mr. Maintenance dude yet. I want to meet him." And just with that 'one' requirement, Steven was already groaning. It wouldn't stop his father speaking. "I know you want to see him, but this is still a new place. I don't know the majority of the people here, and the same rules apply up here as it did down on Earth. Don't go too far from the house, and don't go into anyone's house unless I meet them and their parents. Don't do something that can land you in trouble."

James wanted Steven to feel comfortable, but he also knew that people were crazy, and that not all people were what they seemed. He was sure that everyone was vetted to come on this mission in the first place, they'd not let on anyone outright shifty, but his dad instincts kept on roaring on like the father lion he was.

"Dad, I know! Ugh!" Steven wanted to roll his eyes and almost caught himself doing it, but he also didn't want to be in trouble. "Today I just want to situate my room the way I want to and *maybe* wander around later."

However, his father knew a thing or two about being a child and especially one excited for new friends, so even if Steven were a good kid, he was still a kid, and a very outgoing one at that. Just staying in his room and wandering casually? Not a chance. "This is not the place to be devious, Steven. Don't try to get away with things. The moment you break my trust on this ship is the moment I will shut you down on everything. You don't want to have the most boring life ever. Trust me." James grinned and cocked an eyebrow.

"I won't give you a reason to be mad. This is mom's ship. It would be wrong if her own son was bad." A pause. The tone changed too quickly, faster than James would've imagined it. "...people wouldn't believe that she could keep her ship in check if she can't even keep her own son in check." He whispered.

It took what the man could not to let a soft 'oh' escape his mouth, a deflated sound. James let his shoulders fall from squarer, and a genuine sympathy– sadness - crept into his eyes. But, it could only stay a moment, as he recognized the role they upheld on this mission once more– as if he had ever forgotten. In public eyes, they were leaders of a new generation, not just a father and a son, nor his wife, his mother, just that. So they sat down in the middle of their conversation and talked like they were friends discussing things of great importance.

"I appreciate that, and respect that... just be cautious. What we do reflects like a mirror except we are not the only ones in the reflection," James knew this was probably overkill, but he wanted his son to understand how important this was. He sighed and continued. "I know it isn't fair that you are held to a higher standard than most, but it is important. We are all important now to the entire world. Can you be okay with that higher standard?"

"Dad," Steven began. "You act like this is new. Mom was always going to be amazing. So we both have to be as good as she is. This isn't a chore like doing the dishes. It is really cool to be here. I know what to do."

James thought what an amazing person Steven was becoming, smiling– and then faltered again. Returned now entirely to his nature as a father, not a man of his duty, he couldn't help thinking about how mature it was of his son to make that kind of statement. Admirable, but, was Steven growing up too fast...? With the type of work he and Alyssa were in, Steven was forced to think about the type of things that kids, well... normally wouldn't. After all, they were about to go into space. Into the unknown. That was scary, even for adults. So much to think about– and here Steven was, managing it all with thoughtful expression. Taking it all in stride.

"Thanks kid. I am really proud of you. You are taking this better than most adults would have. Makes me proud to call you my son." And still, despite feeling pride, the worry that he was forcing his son to bypass his youth held strong in the back of his mind. "Hey," he called out. "How about you go take a shower? You smell like an Earth boy."

There was a slight twitch of his nose, with Steven giving his arm a sniff, before nodding to his father bashfully. Immediately after, he went to take a shower and even made his bed. He took his time in making his room just the way he wanted to. The fatigue hit quickly. He would have been tired even if he had slept well, which he hadn't. And even though Steven planned his day out, everything seemed to happen too quickly.

After he finished all the things he could have possibly done in his room, Steven nonetheless went out to venture. He didn't want to go too far; he had no intention of angering his father so soon after landing, and he was still tired. The boy wandered down a couple of halls on all three sides aimlessly, but slow and deliberately, like he considered these his backyard. He paid attention to everything he saw. And while he was on his walkabout, he wondered about things– like if there were potted plants on a spaceport. His mom used to have to tater totter planters. Terra gotta? Tockeria? He couldn't remember what they were. But still she had big plants back on Earth.

He had wondered and wandered a fair amount of time, and came back around to see his father strolling up the area, decidedly at the end of his venture.

"Hey kid!" James said, smiling.

"Hey dad! I'm glad you are here. I was wondering about what those plant bowl things were called?"

A blink; raise of his brows. "What?"

"Those ones mom had the plants in."

"Oh. Uh. Terracotta, I believe."

"Oh. Right." Okay, well, he had his answer. "How was your day?" Steven continued.

"It was actually okay! I was able to start scheduling clients starting Monday, so that is good. But how was your day?" They kept walking together, James having his hand on his son's back.

"Oh yeah– it was a good day. I was just wondering if people could have plants here, and then I couldn't think of the word for those things mom had. The stickycottas."

James laughed. "I'll admit, I'm not sure I can answer the other question, but, I'm sure we can ask someone. Actually– you want to find Mr. Maintenance man now? We have some time before mom comes home."

"Oh yeah! I am so excited to see if he found any good spots. Can we go?" Steven asked, sounding more excited and awake than before.

"Yeah, of course. Let me put my stuff up real quick, and then we can head out. Oh— while I am putting my stuff up, grab three water bottles in case we wander for a while." He requested as they reached the door.

"Good idea!" Steven said with even more enthusiasm.

After they reconvened, Steven was grinning somehow wider. "Ready to go?" He asked his dad.

"Sure am. I think it would be nice if we did walk around a bit. We do have to be home by dinner time, though." James shut the door behind them.

They walked side by side. Steven lengthened his steps and James shortened his and they were in sync. Steven had already been in this area, but James took the time to look around him in wonder. When they hit the great expanse that Steven had found yesterday, James let out a *'woah'* beneath his breath and turned around completely to look all around. Steven smiled knowing that he did the same thing yesterday.

"It is so..." James started.

"Big." Steven finished.

"Yeah. That." James said, still looking around him.

"It is also very clean."

"Well, now son. It is now, and hopefully it will stay clean, but it will also be home soon. You'll see." James put his hand on his son's back again and smiled.

"How do you know?" Steven asked, looking up.

"Well, think of it like," James continued to look up and around while thinking of the proper thing to say, "It's a new pair of shoes. You take them out of the box and they are so very clean and smell like leather. They are almost too shiny and you know that they have never been worn before. But after a while, you break them in, right?"

"That makes sense." Steven smiled.

"All we can do is make it our home. This is an adventure of a lifetime, son."

"Yeah it is."

They stood together, looked around them in awe when someone started walking up behind them.

"Hello, friend!" The voice called out from behind.

"Hi! This is my dad!" Steven chimed in response.

The man walked up and reached out his hand to James.

"Hi, I'm Mario. Nice to meet you." The maintenance man said to James.

"James." He replied, returning the shake. "I heard you kept my son company a short while yesterday while he was thinking. He is a very curious young man."

"I want to know about everything you do!" Steven jumped a little bit.

Mario chuckled and began talking to the two for quite a while. He told them about how he would make sure there was no debris on the ship. He would also be in charge of the cosmetics, even to paint the rafters when they needed it.

110

"Rafters?" James asked, impressed by the workload that he had, and the pride he had in it.

"Well, I do also make sure the windows are clean. So, the rafters get walked on. Eventually they will need to be repainted."

"Can we go up to the—"

"No." James cut in before Steven could finish his sentence.

"Ah— no," Mario also started to say, looking a bit shy as he spoke. The hint of his Spanish accent had started to show through the longer he enjoyed speaking with them.

A dramatic huff followed them both, "Boo..."

"But Steven did mention that you would be on the hunt for the best room on the ship." James redirected.

"Ah yes, I have the perfect room. We will have to be quick, though. I have a lot to do yet, and I don't want to be late for dinner."

They agreed to be quick and Mario led the way. The conversation was easy and light between the three as they walked. Mario talked to James about what they each did, and then asked Steven about his interests. Then Steven and James asked about things that made them curious. Even about if people had plants— some people did, in fact, have them. It wasn't as easy to maintain, as they had to have special lights on to keep them growing, but it was possible. Then, James asked about Mario's family. All the while they continued to walk around a non-descript area. They passed a door that looked heavy and had a security pad next to it.

"What is that?" Steven asked, stopping to consider the heavy door.

"You know?" Mario hesitated, thoughtful for a moment. "I don't know. I don't have the codes for that one."

111

"Hmm. I don't either." Dull input. "Restricted access." James spoke mildly. Strange, for a moment; neither could tell what he felt about that. Then, in a slow, stilted turn, the man made a ghastly face and raised his hands to wiggle each finger spookily. Steven laughed and Mario smiled.

"It's probably for the bosses." Mario remarked, breaking the moment and leading the group once more.

Along the way, they stopped again for something seemingly important to Mario.

One of the walls opened up to show the wires underneath. There seemed to be cables and wires that wrapped into great vines.

"We are working on fixing the panel." Mario explained. "But down here is where I want to show you. It is really cool."

The taller man led them to an area that looked somehow more intense when the doors shut behind them.

Steven quietly took his father's hand, but then the entire room became bright. It was a shooting star. As the burning star passed, there didn't seem to be any walls. There was a large area to walk on, but the entire room, above, below and around was glass.

Steven breathed out in awe. "Dad?"

"Wow..." James couldn't help being just as woeful as the child. He'd never imagined something like this, let alone witnessing it firsthand. He felt like he was standing with the stars themselves, out there in the endless expanse.

"Right?" Mario looked around with a smile. "I found this room and took too long just hanging out and looking out."

"I can so understand that." James muttered, completely captivated; the sight around him was breathtaking.

After a while, James looked at his watch and decided that they finally needed to get going. The vista was so amazing that it overshadowed everything else– time seemed to stand still. But that wasn't the actuality, it was going on with or without them. So, they had to move on. Steven asked if he could go with Mario and hang out. Mario didn't mind, but James quickly ended that idea. He told his son they had things to do with Alyssa. They had to get home and take a shower, and dinner was very important, as much fun as exploring was.

"Hey, Mario. Thank you for this. And listen, if you don't or can't have Steven with you, we both understand, but I am okay if he does hang out with you a bit." He almost left it at that. But, Steven's interest was piqued, almost like this was a hopeful loophole to go home a little later. Clearing his throat, he was quick to add, "–just not right now."

"Thank you. I appreciate that. I don't know about having guests *all* the time, but every so often I would be honored." Mario shook the father's hand.

Even if he couldn't stay, Steven smiled anyway. "Thank you for showing me this cool room with my dad, Mr. Mario."

CHAPTER NINE

One year later.

"Hey, Mr. Preston!" Steven called out while he was running down the long corridor.

"Heya Steven, where are you off to in such a rush?" The man named Mr. Preston called back. This fellow was short, but stout, with black hair and dark skin that was complemented by his olive colored shirt. Everyone wore basically the same thing, though the slight differences were chosen by each person by way of color and mild variation of style. Some had a V-neck, some had a crew cut, so on.

"I'm so late!" Steven hollered while looking over his shoulder. He continued down the hall and took a right, a light blue blur of a boy– that was *his* shirt of choice.

Steven knew the ship as well as the back of his hand by now. Well, mostly. He still didn't know any other shortcuts, only this one. He spurred on, feeling the burn in his legs and pushing to make them go even faster. His shoes were squeaking slightly at the twists and turns. He had spent too much time reading in the room with the big comfortable sofas. But it had to do with

115

important ambitions, in fairness! He was determined to read every book on the ship before they were done here. It was his mission, as much as being on the actual one they were on. Which, sprinting at the moment, he was trying to take part in the latter.

He was so upset he was late– would this reflect poorly on him? A part of him was sure it did.

It aside, things were going really well otherwise. His dad had a great client list at the gym, and he loved getting people in shape and keeping them healthy. Steven had talked to his dad just last week about maybe starting to lift weights too. They had decided... to not lift quite yet, as a kid he still had a bit of a way to go. But his dad would teach him how to get in shape, and make him faster. Then when he grew a little bit more, they could work the iron.

That was important, too: three hundred and sixty five days aboard this ship, Steven thought of this mission as his own, not his mother's anymore. He planned to be the one to know everything, and be in everything. He was especially keen on faces, names– he was friends with all of them as of not so long ago, as far as he could count.

Finally, he reached his destination.

The greeting was crass: "You're late, Steven!" A boy named Randall shouted, arms crossed. A feigned glare followed Steven while he slowed down.

"I know! I got caught up!" Steven swung his backpack from behind him and retrieved a glove from inside the front pocket.

The other boy grinned mockingly. "*Yeah right*, you were probably reading." Still, he tossed the ball to him in earnest, backing up a few feet every throw. He waited of course until he was able to return fire– a fair chance for a fair game.

"Whatever. The book is good." Steven smirked as he caught the toss. He pelted it back hard enough to make his friend inhale sharply.

"It's alright, but— *man* don't throw the ball too hard. Being a nerd isn't that bad of a thing!"

He didn't throw the ball like Steven did, but he didn't flop it over either. Being a kid that was a good five or six inches taller than Steven, he knew he could send it harder, but he never used this against his best friend. They understood each other. Plus, they played enough to fit a routine to their motions. They had practice every odd day, as long as they didn't have too much homework; when they had testing weeks, they met up every Tuesday and Thursday.

In the nature of good sportsmanship, Steven shouted: "Nice throw!" And threw in their usual spirit. He wasn't going to bring a bad mood to this just because he was late, that was for sure— it was a perfect day for a good game with all the right energy. Steven was *sure* he'd be 'reporting back' to dad in his best shape.

All the while her son was having the time of his life, Alyssa had a list going in his head that she needed to write down somewhere, anywhere; gruelingly exhaustive as the type of things that needed to be *exactly* kept track of. All there were was problems on this ship. She got some more coffee and an apple from the break room table in the back. Everyone had been up for hours and hours... The plan was to work on each individually, if they could afford it.

First, the main server went out. *Ker-plow.* Just decided to go completely dark. Alyssa sent Joe to work on it in the designate room. Without physical input, the issue wasn't easily resolved. Instead, they stayed unsure if it really was their server, or the satellite; few things gave any clear indication. And there was no hopping from one satellite to another like people would a phone tower because, *well*, satellites had to be in line of sight. Though, she believed Joe had it covered... The upload screen was on and waiting.

Everything seemed to be on the fritz today.

Ah, but *really*, it started at the beginning of the week. Monday, everything seemed to happen at the same time.

She hadn't even had much time to think about the fact that the docking bay door cracked open for no apparent reason. A cracked door in space... nothing could go wrong with *that* outcome, right? Oh, and let them not forget the generator that overheated, or the number of computers going haywire in rapid succession.

Alyssa stared out blankly, seeing only the problems in front of her, and not the room. She saw them all together, and then she saw them individually. As soon as she could talk to her supervisors down on Earth Point, which was what she was stuck waiting on, she wanted the 'go ahead' to essentially remake her entire system from scratch. This wasn't sustaining them as it was now, it... *the problems* needed solutions, and quickly.

It had only been a year.

In this one year, she needed to rebuild something as essential as the docking bay door because it was breaking down.

Eventually, too much pressure would build and cause some real damage if it wasn't fixed. It could get catastrophic.

But everyone, and everything was okay right now. She handled it like she did everything else. One at a time.

"Hey, Mary?" The woman called out, still distracted and staring out, "I will be in my office trying to call Earth Point. Uhm," She paused before looking over to Mary directly, "Can you get Manny on the bay door with Cody and tell, ah, tell Joe I want the server report whenever he has time, but that it definitely needs to be done. Also..." She took another break, and inhaled... A long, controlled breath, never wavering or whistling as she did the best she could to hide exasperation... "Order us all some food. Maybe some pizza? Ask what everyone wants, though it's been a long time since we had a pizza day. We are going to be here for, it's— it's going to be a while."

"Okay, great, will do, but Captain?"

"Yeah?"

"Breathe. You're doing great. Really."

"Oh Mary. Thank you. Today has been a lot, right? But I am going to see what can be done from the people down below." Alyssa stirred her coffee.

Mary began to walk back towards her station.

Though, her captain spoke before she was too far, "Oh, and Mary?"

"Yeah?" The other answered as she turned her head back.

"I'm sorry for bombarding you. I just feel like I am running on an incline today."

"Run, Captain, run." Mary chuckled softly in her best Forrest Gump impression.

"Hardy har. You're lucky my husband loves him some old school army movies."

They exchanged lighthearted smiles, a quiet laugh, before Mary left, closing the door behind her.

While Alyssa was saving the world, James was working just as hard.

"Run, Billy, run!" The man yelled out while rushing around the treadmill at break-neck speed. He cheered the other person on, before exclaiming, "*Yes!*" He leapt with enthusiasm. "Your best time yet!" Just as soon as it was clocked in, James offered a high five, all throughout grinning widely.

Billy was a big man. Not fat, more burley. He was just big all around, both wide and tall. He had gained a little weight while he had been getting comfortable in his new life. It seemed to be human nature to slack off of working out while getting their life in order in a new environment. Changing your life sometimes means needing to adjust and find an equilibrium in your new area. And one could definitely count moving to space as a major upset of one's normal life.

"Okay you go three more minutes. Walk it out, my man." James walked towards the towels, picked one up and threw it at him.

James had three people cancel on him today. It was odd, but he was looking forward to some house cleaning for both the gym and the house. On the ship, there were things that could mostly be handled for you, like food. You could go to the mess hall and eat what was served, and never have to worry about the dishes. You had an option to cook, though, but the galley kitchen didn't offer much by way of space enough to cook. There was also laundry. Certain people handled the laundry for the ship, but you had to gather it and sign it in and out. Racks were collected at night when you put the outside at night when you were done. It was a fantastic process. Though sometimes, you just liked doing things on your own.

James pondered some more about the state of the ship. After his last client left, he switched from work mode to clean mode. The mess hall was a great thing, but man he missed grilling steaks sometimes. James loved the American dream. And besides the missing of steaks, he felt like he was living the dream up. Maybe he could request a smokeless grill.

The thoughts helped him clean and sanitize his work space and then he began to walk home. Mentally he made a list of the things he wanted to get done before Alyssa got home. It was going to be a late night for her again. Lately, there seemed to be a lot of problems and she was beginning to feel the effects of it. So the plan was to feed Steven early, get him to bed on time tonight, and then make Alyssa a bath so she can relax.

Steven had been sneaking out at night to read. James was going to bring it up at dinner tonight if he didn't confess first. Usually his son would bring things up when they were alone and settled, and lately there hadn't been much of that. During the times when it was just the two of them, or the boy with his mother, he would talk about different things. His fears, the things he wants and anything else he felt merited a conversation.

James was able to get through his whole to-do list relatively fast. He had deep cleaned, taken the laundry. For the first time, in a long time, he realized how quiet it was. With Alyssa at work and Steven at school, it reminded him of when he had days off during the week on Earth. He did the best thing he could do now that his list was done and he still had hours. He took a nap.

Beep. Beep. Beep. Beep. Beep. Beep.

Alyssa sat at her desk with her cheeks between both of her palms. She had been staring at the phone, finding that she wasn't really hearing the alerts anymore. The captain had been calling off and on all morning to no avail– at this point, every time she called, she zoned out to wait for the voicemail again. It was nearing levels of 'hopeless endeavor'. It could not be stressed enough as important– to not only her life, but to dozens of people aboard. Especially as these were people who trusted her to keep them safe. If... if this didn't go through, what was her next plan of action...?

How long had it been, now? Alyssa couldn't count the difference between tens of minutes and twos of hours.

Then– "Sergeant Armes?" *An answer!*

"This is Captain Burns with Space Port." Alyssa barked, jolting straight up. She hadn't really expected anyone to pick up at all at that point, but this was a huge relief.

"Hey! Well this is a surprise, Captain. We aren't scheduled for communications until next week. Everything okay?" The words were nice enough, though he sounded disgruntled, as if calling from the SpacePort early was a total inconvenience.

"Crew and all on board are fine, but we are having a lot of computer and internal problems. I have been attempting to send out communications all day to try and let you guys know, and it hasn't been going through, hence the call. I am wondering if you guys can do a deep scan on your end. We have technical problems beyond that of the norm. It is not just one or two things. Everything seems to be acting strange." She was playing with her pen as she spoke, holding it loosely and letting it make little marks on her notepad.

"Oh." To her concerns, this was... all he said, *'oh'*.

There was a pause from the other line that took so long, Alyssa stopped fidgeting altogether to look at the phone.

"...Hello?"

"Hmm." The Sergeant cleared his throat. "I'm on the line, Captain."

"Well, can you help me with diagnostics?" She was slightly unnerved that he seemed so... *unconcerned* with it.

"I'll look into it. But I will have to get back to you. I–"

Alyssa's sigh was soft, but it was heard from both sides.

The man on the phone continued. "I really don't know what to tell you. But you know, if I find something, I will let you know."

"Okay...?" Alyssa knew something was wrong but the brick wall that was just put up before her was... really tall, and firmly put in place. Why weren't they confused? Or worried? Even a little? "Well, I'll have my guys continue to troubleshoot it here and see what we can come up with, too."

"Roger that."

The line went dead and it left Alyssa sitting there wondering what exactly happened on that call. It was so cold, so uninformative, the captain couldn't help wondering if this really only wasted time. He didn't want a list of programs infected, nor did he even seem concerned. Everything about that whole interaction was just too weird. Maybe he didn't think he needed a list if they were doing a deep scan? Her mind was racing. What was that all about? Did she say something to upset him? It couldn't have just been because she rang early, could it?

"Captain?" Mary knocked and cracked the door open.

Alyssa was rubbing her face with her hands as she answered. "Come in."

"I'm sorry to disturb you. I saw the line went off. Uh– we have another problem."

Mary came in and sat down, managing to look less anxious than Alyssa felt. She had a computer tablet and a clipboard in her hands. Today, they started using both because the primary systems just weren't reliable enough– which wasn't something she wanted to think about while living on one, big *machine*. But, it was nonetheless necessary to make sure they had hard copies of everything in case the whole thing shut down.

"Hit me." Alyssa nodded, sitting back in her chair.

"Actually– it is... two problems. Well, one problem creating another problem. There is an influx of pressure. Which... the pressure isn't enough to harm the ship, and it seems to happen in only random places. For some reason, it is making the cameras look warped." Mary handed the tablet and clipboard over.

Alyssa took what was offered and saw that the security images looked as though they were being held underwater. "Okay...?" She flipped through the 'reel's again, "Just confirming there are two issues, one creating another. One is the pressure and the second is the cameras. Can we confirm that the cameras are warped because of the pressure?"

"I, well– see," finally– *no, that's not good*- Mary seemed at as much of a loss as Alyssa was. "That is *the* problem. We can't get a proper reading on how much pressure there is. Only that there is pressure. It isn't showing up on any of our diagnostics. But it is enough to mess with the system itself and tell us there is a problem. We can't get a read on how this is happening, either. Pressure is assumed because... It isn't one camera at a time. It is all of them at once."

"Well..." Alyssa looked around the office trying to put the pieces together, and... she was failing. "What the hell? We just checked the pressure sensors and they *are* working. This doesn't make sense."

"No, it doesn't. But it is clearly there."

They took a moment to ponder over this mystery, before simultaneously, the both of them stood to their feet, inconsistently shuffling to grab what they needed from the desks. It was an unspoken decision to make their way to the bridge entrance. Unsurprisingly, Paul was waiting for the both of them with a stack of papers and his own tablet.

"I was just coming to your office." He had only caught his breath to speak, doubling back to walk with them.

"Talk to me, Paul." Alyssa demanded while still keeping her pace to the center console.

"It's all normal. I mean— it's obviously *not* normal, but the *readings* are normal. There's clearly something wrong, but the calculations are all what they should be. I ran them every way I could. I even checked them by hand!" Paul started flashing random pieces of papers as though either of the two could read his handwriting and mathematical mess as fast as he could.

Alyssa stopped and looked out one of the large glass windows— to the expanse of space beyond them.

Her eyes traced the stars, seeking familiar patterns. It was unlikely. They could map and mark them, but she was not on

Earth's ground, not a child looking up as a distant witness. This was not what she knew as a girl, the viewpoint she trained herself to then. And, this is not what she knew as a captain. Too much at once, too little information... there needed to be a reach. Something– anything. Whatever it took to draw the lines of Orion's belt back together; whatever it took to make sense of the failing instruments. "...Maybe we need to consider an outside source?"

"Like?" Paul asked, pausing in his step.

"Are there any solar anomalies? Flares– or anything external that could be a factor? Maybe the right kind of flare at a certain level can throw things off just enough?" The captain asked while looking through data on her own screen now that she hit her center console. She almost sounded desperate.

"Negative, Captain. There are some flares being detected from the star, and yes, that can account for some of the connectivity issues we have been having, but there is no way it is creating enough energy to pull all of it; we're in a solar minimum right now. And even if we weren't– it wouldn't push enough pressure to register on our system like this, not unnoticed. I can't imagine that." He answered while looking at more math papers that were once in order.

"Okay. Well, I talked to Earth Point, and they... weren't very helpful, but Sergeant Armes said he would look into it. Meanwhile, what can we do here?"

"Captain!" Mary called out a couple of consoles back, "Speak of the devil! It's Earth Point."

"Okay, everyone be quiet. Mary, throw him on speaker." Alyssa called out as she sat down, more than ready for some actual answers.

"Sergeant Armes. This is Captain Burns. You are on deck speaker. There are a few more anomalies that have sprung up since we talked previously. Do you have any updates on your side?" She braced her hands on her desk, if it could be called that. 'Braced' was a good word for the way her body was rigid, forcefully balanced. Alyssa was hoping for answers and solutions. Her usual standard operating procedure was to pace, but that wasn't going to work if she needed access to something. She did not want to be seen fiddling with her hands. Hence the bracing.

"Okay, so I have good news, and not so good news. Which do you want first?" Armes' booming voice came across loud on the comms. He still sounded annoyed– like this was a matter not worth his time.

"Lay it all out, sir." Alyssa inhaled, raising her hands in a frustrated manner– she was nearly ready to smack the desk. She really didn't have the patience for the good news, bad news thing. She just wanted to know where she stood. *What was happening?*

"Look, long story short is that there is an explanation, but it won't make you any *happier*, because there is nothing we can do." He sounded tired. It was offensive.

That didn't answer anything. "And the longer version of that would be?" The irritation was seeping into her tone.

"The boys on the long range scope found a star that went supernova. This usually only affects readings in the beginning of our deep space area. However, its blast trajectory mixed with the location of the star which mixed with minerals that we can't yet identify has made the solar flares in the sun magnify. A lot."

There was a silence in the room that was deafening. Sergeant Armes continued, "This is creating a magnetic pull and a reaction six to eight times worse than the usual pull. Again, this doesn't usually mean anything for us. We are equipped to handle normal ranges up to four times. But this is worse. Another reason this is affecting you is because you are on top of the atmosphere. Your magnetism levels and radiation levels are beyond that of what they should be." It sounded as though he was being given the details on an index card to read off of.

Alyssa considered everything for a moment. Though no one could exactly predict a supernova, they did have scenario runs for it. The launch had been made in consideration of their star's weather when it was supposed to be most stable. This was... unprecedented. Perhaps it was just the placement and chemical make-up with the minerals that went beyond their training? It was assumed that there are many different make-ups in the stars. They haven't found all the minerals in the vast space, after all. But they couldn't run diagnostics with unknown data.

"Okay." She breathed: in, and out. She stood up... and started pacing. Her natural protocol. "What do we do?"

"We don't know. We can't predict what will happen because the sparks it is giving off isn't anything we have ever seen before. What we think will happen, based on the information we do have: sensors and readings will continue to malfunction for a bit. Then the flares will dissipate and everything should go back to normal." He cleared his throat and there was a clank, as though he was drinking from a glass. "Now, from what we can tell, the ship should still run like normal and nothing seriously wrong should happen. We suggest not causing a panic with the families on board."

"Are we going to continue looking for further possibilities? I am not sure that explanation accounts for all of the problems we have been exper–" Before more could be said, Alyssa was cut off by the voice on the other end.

"I wouldn't waste the resources. The data is what it is. Which isn't a lot, but our sources here suggest that could be the cause of all your problems. There is no imminent danger or trouble with your facilities. We are running projection theories and will send you a full disclosure report in a couple days." And that was that. It. Final. Brushed off, really, but what could she do?

"Copy that." She muttered, crossing her arms over her chest and attempting to hold in some comments that would very likely get her in trouble.

The line disconnected and Alyssa felt like the information she was given was just... wrong. She had to address the crew, though, and she had her orders, though they didn't make a lick of sense. Something was off.

"Okay, guys. I am pretty sure we are all in the same frame of mind. The one thing that I can rationally say is preserving our resources. My thoughts are holding off on anything that doesn't need immediate work until that report gets to us. The things that need to be taken care of include the bay doors, though. So let's work on the physical stuff for now and wait for more info."

"Captain," Joe called as he stood up, "With all due respect, none of what the Sergeant said makes sense to me."

"Yeah," came an exasperated groan. "I am inclined to agree with you. But we still have no other options, given the information that we have. We've got to wait for their report."

"I have studied these types of anomalies and I am absolutely sure this isn't the cause. Supernovas burst at a specific rate.

That is why they are called supernovas, in part; they've been studied for decades. We simply wouldn't go without catching one of *this* magnitude, and no one element – *mineral* – could impact or increase anything to this degree. Not without there being more to it; the factors would be very different, and still, we'd have recorded that *somewhere*, even if just simulated. In all preparations for launch, would this not have been noted? How could this just *happen?*" Joe countered, looking incredibly self assured. It was all she needed to know that they needed to try to handle this themselves. After all, if Joe knew this information, so did Space Port– and they still said that garbage.

"I agree. Then, let's run our own report, and we can compare to theirs. We don't have what they have on Earth Point, though we are closer than they are and have all the brilliant minds we need to do this. Let me know what you need, and I'll see you get it."

CHAPTER TEN

Freedom is a *relative* term.

Justin may have been released from the holding cell at One World United, but he wasn't free. He was being sent to Earth Point until further notice. Likely, he was going to be kept there until the end of the mission. Of course. What was the saying? Out of the frying pan and into the fire.

He wasn't doing well. After everything? No, not at all. Because besides the obvious, his thoughts, which began coherently, tangled themselves into an unrecognizable mess.

He knew why they – whoever 'they' were – released him. It couldn't be more blatant: Earth Point.

It would be close to time for the launch, and they would likely need him. *Need*– no, *want* what they could get out of him. His math didn't lie; that was something he knew Jeff and him both felt. Their math was never disingenuous, and Justin was the only one who knew some of the formulas they may need at any given time. After all, *he* was the one that *created* them.

The man spent over a year in a cell for telling the truth.

Really, he wished he had done things differently. Not that he regretted what he did. Not a chance in hell. Rather, how

he executed it, the time he did it. Waiting only hours longer...
a private audience could have been arranged with the family,
among other things. There were simpler answers that didn't end
this way. Justin was supposed to be one of the people responsible
for their mission, after all. Trust could have easily been gained,
and then, he'd say his piece about the right things, sorts that
would have actually helped them escape their doom. Instead,
he'd just made everyone think he had lost it.

Society thought he was crazy, or so he'd been told.

Despite this, he was going to work on being knee deep in the
project and everything that came with it. Alex was going to pick
him up soon. Wished he had a pack of smokes. In their little
work group, only Jeff smoked, but after a year in jail, Justin had
picked up the habit. Now that he thought about it, perhaps he
even called them smokes because Jeff had done first. However...
shared habits were never talked about in this place. It could be
dangerous if it wasn't considered proactive. God forbid Jeff get
accused of collaboration in this 'conspiracy'– he didn't need him
here to tell him how things changed.

And Jeff just wasn't the man handlers imagined they were
shaping Justin out to be.

Thanks to his time working as such a 'well-liked' personnel
of the One World United, he knew the right things to say and
do, and they'd spoon-feed correct answers for responses deemed
wrong. Even if acting out could be called a crisis, he knew he
would eventually be led back into the folds. They needed him—
he kept being reminded of that, somehow. Or led to think it.
The entire time he had been held and given the 'therapy' that
councilors delivered to reprogram his mind – fix those *"paranoid*

delusions" – he was just learning how to best slip through the cracks they created.

The biggest challenge he had faced in captivity was the day everyone was forced to watch the launch. The entire detention center had been gathered in the mess hall, and a massive projection screen broadcasted the live event. And although the place had been packed, Justin could see certain doctors and orderlies watching his facial reactions. There was a pointed range of emotions that monitored him for, and that he was truly feeling– but he only allowed the part of him that felt genuine excitement to show.

After the launch, there was an extensive conversation with Justin and 'the therapist'. Presumably, the meeting was to ensure that the man was mentally stable after the launch, though that was only part of it. They wanted to see if he would break again after seeing the family.

"How are you?" Justin's therapist asked.

He never did learn her name. She was short. Really short. Had to be borderline to the height considered for dwarfism, and he wasn't being rude about it. The woman was really just that small. She had long brown hair and cold brown eyes. The vest she wore made her seem even smaller. Maybe her name was Sarah? Samantha? Sybil? Didn't matter. He only ever called her ma'am. It was his silent protest; to be able to purposefully not know her. He couldn't really even talk to her. It wasn't like he could tell her about the things that he did for One World United, or what he learned.

"I'm fine."

"I noticed that during the flight you were a little on edge." She leaned forward a bit in her chair.

133

"Of course I was." Justin snapped.

He reigned himself in. The man knew he would never get out of here if he didn't. He had learned how to channel his emotions and answer in a way that they deemed more appropriate. His hands rubbed over his face roughly, and then he sighed.

"Look. I helped make all of that happen. I wasn't on edge because of the crazy shit— *err, shit.* I mean to say stuff. The stuff I said before."

"Okay. Do you want to tell me why you were on edge, then?"

Like he had a choice? He looked at her, with her legs crossed at the knees and her hands clasped over them, as if she was waiting for a toddler to answer her. He looked at his hands. He just needed to tell them what they wanted to know, then he could go back to his cell and brood.

"Sure. I'm just trying to find the words. See, I made that happen, as I said... Well, I played a part, anyway. And because I cracked wide open, I wasn't there to make sure it went okay. So I was on edge, because..." He trailed off and looked out the window.

"Because you should have been there." She finished for him.

"Yeah." He sighed. "Yeah, exactly."

"You okay?" Alex muttered, not looking at the other, but feeling the tension in the air. The memory from the cell was still permeating the air they were breathing.

"Yeah," Justin answered quietly. "I'm just nervous. We both know they won't be happy to see me."

"It'll be okay." His friend responded, not sounding so sure himself.

Justin stayed silent. Alex's words and emotions seemed to be a mirror of his own. They drove for hours. They stopped only to fuel up and eat. Finally, in the dead of night, the lights of Earth Point began to show itself.

"Here we go again." Justin said to himself after they parked and he got out of the car.

It may have been midnight outside, but you wouldn't have been able to tell on the inside. There were people milling about and working. Night crew were on deck and in full force.

Alex made his way towards the elevator, "We are meeting over here."

"Who is it we are meeting, exactly?" Justin figured he probably should have asked sooner, but really, it didn't matter.

The good thing about the jail term, if there *were* a good thing, was that Justin had been pretty good at knowing when his emotions were about to be shown. He didn't know how they got away with doing what they did in there, though as the ruling power of the world, there really wasn't anyone to stop them. He also knew he probably wouldn't be able to tell anyone anyway.

"Clinton." Alex finally spoke. It was a firm, singular word.

It took a moment for that to sink in. What had he asked...? It-*oh*. It was who they were meeting. Ah. Clinton. It made sense. Justin should have known. Somehow it still seemed to catch him off guard. Clinton acted as a Human Resources Director and was the official liaison between the workers and the governmental powers at be. If it was just Clinton, Justin had to assume things were not going to go well for him. Clinton had always been a yes man. Meaning, he usually only did what would make him look good to the higher ups.

"Hey guys," came a voice from the other end of the hall when the elevators opened.

There was a casual jest from his co-worker. "Stalking us, Clinton?"

It was good to see that some things didn't change. Alex never did like Clinton, thought that he acted more as a liaison than that man ever did. *He* actually fought for the workers.

"Just got here myself, actually. Was headed towards the office when I heard the elevator ding."

"Mm." Alex hummed by way of response. "You remember Justin, don't you?"

It was sarcastic, but Clinton didn't seem to notice.

"Of course!" He held out his hand to shake it. "Good to see you again." He didn't look like he meant it, but it was a decent attempt.

"Yup. You too." Justin answered back in likened fashion, shaking his hand.

Once they were inside the room, the three took their seats. Alex had stretched his back so hard on the chair, a pop and crack sounded off. Justin stifled a yawn as Clinton set his briefcase down, opened it, and pulled out a file. Wasn't just a paper. Never was. He knew it was personally relevant– didn't take much effort to confirm his suspicions.

"I want to apologize for you being kept in there for so long." Clinton remarked simply as he opened that file and started shuffling through papers.

"Jail." Justin retorted shortly. "I was in jail. For a year." He knew how to keep himself in check, but he didn't feel like hiding all of what he felt. He was allowed to show some of his displeasure at the male sitting before him. It was one of those cracks small enough the organization could not illuminate it... of the things he learned his right to exercise.

"That was out of my control." Clinton replied calmly, putting the documents down and staring at him.

"Was it?" Rhetorical. It wasn't a question. And Clinton knew it.

"Okay guys," Alex raised his hands. "Let's take it down a notch. Justin, we can't change what happened anymore. We've got to move on if we are going to do anything about it now."

The statement was confusing, but Justin let it slide. Everyone was tired, and he got the gist.

"Yeah. You're right. Okay, then." He said leaning back in his chair and looking right at Clinton, "What now?"

"Technically," Clinton began, looking at Alex because he was unable to meet Justin's eyes just yet, "Justin didn't break any of our Non-Disclosure Agreements. Nor did he break any of our policies. As far as we are concerned, we can give him the same deal that we gave Jeff and move on." The man glanced back to the writing before him every so often.

"Jeff?" Justin's head snapped towards Alex's so fast his neck cracked. "He got into trouble for seeing me?"

Alex scowled at Clinton. "Yup."

"No." Clinton spoke at the same time, glared back at the man.

"...Well, which is it?" Justin answered more calmly than he thought possible, looking between the two.

"He got slapped with some probation and counseling. That's all." Clinton tried to placate.

"That's all." Justin drawled.

"So that's your deal. Counseling and probation. Besides that, everything goes back to the way it was." Clinton seemed to find whatever paper he had been looking for, and handed it over to Justin with a pen.

"I don't really have a choice, do I?" Justin inquired in a deadpan tone; this was not a question either.

The idea of counseling made him cold. The paper was pulled towards himself and signed sooner than he could linger on the wording. He wouldn't be allowed to actually have a voice regarding any 'deals'. Couldn't afford not to do this. His family suffered while he'd been gone. They had to sell off their property and were struggling to make ends meet.

"Well, that, as they say, is that." Clinton smiled. "We have a house set up for you next to Jeff's. We figured you would want your own place after being cooped up for so long. Go and get some sleep. You can surprise him in the morning."

"Come on." Alex got up, not bothering to push the chair back in. "I'll drive you there."

No one made an effort to shake hands, part like proper working men. There was no one around that would take note of the lack of formalities, and they already shook hands in view of the hallway cameras. The only place they'd ever need to— no need now to feign pleasantries. They took separate elevators and Justin hoped he wouldn't have to see Clinton for a long while.

The way to the house was shorter than expected. Justin thanked Alex again for driving him.

"You going to be okay?" There was genuine worry in Alex's tone as he watched him walk towards the entrance.

"Yeah. I think I might crash out." Justin mumbled. He knocked on the car hood twice and turned to his home.

CHAPTER ELEVEN

Randall wasn't in school.

This was not a good sign, as far as Steven was concerned. He had *also* barely made it to school on time, but Randall hadn't been on those escapades, so it made no sense for *him* to be gone. He kept sneaking out at night, which meant he wasn't sleeping as much as he probably should be. Now... he felt guilty about it. Steven had almost been caught by his father, and because he was almost late, he didn't have time to see that Randall wasn't there before school started.

The boy was sure that his father knew, or at least *suspected* that he had been sneaking out. And– he *knew* he should stop, but lately he felt like he was in a giant fish bowl. At an age of curiosity, Steven craved being able to see the outside of the tank. He didn't know how to explain this to his father or mother, but he also didn't want to be grounded– his very first sleepover was this weekend! Randall was coming over. Pizza was already ordered from the kitchen, and his dad was going to deliver it himself. It would be great.

Maybe Randall was sick? That'd be terrible– they had planned their sleepover weeks ago... and they were also supposed to

141

play catch tomorrow! Sure, they couldn't do it every day, but he knew their habits pretty well by now. Steven wanted to be a professional baseball player, so practice was key. There were no leagues in space – not yet anyway – so he and Randall did the best they could here to 'jumpstart' their careers. This way, compared to down there, they could hold their own in a good game when more people came up. Maybe they'd even be better! It was space, that was *their* turf, and both of them hoped they could be the first generation of pro space baseball players.

Today wasn't a practice day, though. Today, instead, was dedicated to hanging out with his dad, and next to nothing could've topped that. It was... just a bummer that he couldn't be as excited as he wanted to be for it. There was no helping just how anxious he actually was about his best friend. It didn't make any sense. He seemed fine yesterday. Something happened that made it all especially bad– it clung on like a parasite, *"Why would they say that?"*

After school, when Steven went to go meet with his father, he decided to take a route that brought him by Randall's living quarters. There was hesitance– a breath he didn't realize he was holding. He didn't want to intrude on their family... he was just really worried. But, it'd be quick and uninteresting, he told himself. Just a peek.

Drawing in closer to it... there was that feeling you get when you miss the step off a curb. A lurch in his stomach, a foot had fallen but he knew it hadn't. Closely following nevertheless was that sense of something *horribly* wrong, he *must've* tripped. It was something like the moment you have nothing to catch you underfoot, and your mind goes through the awful possibilities of what happens next, from as little as spraining an ankle, to a life-changing break.

Or, nothing at all, and you just so happened to fall on a missed step: embarrassing. That was all there was to it.

No, the boy did not trip at all. Nothing happened. It was just... dread, passing by a house that looked like any other. What Steven now felt himself looking at were the very eyes of the concept of proof– that he had nothing to present, not any evidence of strangeness at all. Still, something was *wrong*. He wanted to swear on it, he felt like he *needed* to. If not for his sake, then what about his friend? But... with nothing to show for...?

He could be reading into things, and didn't want to tell anyone about it. Would they make fun of him? Brush it off?

This was so particular, the kind of thing only immature kids *really* got stuck on. They'd make up ghosts and believe in boogeymen their parents told them about just so they wouldn't be stupid about things. He'd read books like that, plenty of horror and all the same plenty of things about real phenomena; adults scared themselves, too, senselessly.

But... right now, he was caught up in something like that, wasn't he? The boy couldn't let it go, so... he was a child no better than the others in all truthfulness. Would this count as failing his mission...?

It tasted like bitter poison in Steven's throat. He couldn't even make odds and ends of anything.

There was only one thing that could make him feel better– and she happened to round the corner in front of him.

"Mom!" he called out to her while breaking out into a jog. The relief in his body was immediate, his shoulders sagging from a tension so extreme one would've assumed he'd been face to face with much worse than a friend with a cold.

Alyssa's head turned to look back, noticed him, and smiled a playful grin. She sped up in a 'chase me' sort of way and chuckled as she did it.

"Mom! Wait! No, I need you!" He cried, the panic swelling again.

She stopped. "Uh-oh," The mischievous smile stayed as she turned to him. "Is the world ending? It's a good thing we left, then, huh?"

Steven ran the rest of the way and hugged her so tightly that she frowned and pulled him back. Something was off– wrong, something was wrong. He was scared. At least, that was how it seemed he felt.

"Hey. Are you okay? What happened?" His mother pulled him close reassuringly, rubbing his arms.

"Mom, I'm not playing right now." That was all he said, all the anxiety that he had been holding seemed to be bleeding from him now that he was finally able to let it.

Alyssa squatted down so she could see him eye to eye. "I can see that. Why don't you tell me what's going on, yeah?"

"Well, you see, I was getting excited. You know, because we have that play date set up with Randall, remember?" He spoke very quickly and was shifting from side to side nervously.

She nodded her head, "Yes, I remember. Fridays, after school. Continue."

"He uh–" Steven rubbed the back of his head with a hand. A habit he got from his father. "He didn't come to school today."

"I'm not seeing how this is an emergency, son."

"Mom! You don't understand! *Ugh–*" He barked, clearly upset. Now all too like his mother, the boy was pacing back and forth.

"I think we need a five second breather. I don't know what you're talking about and I need you to calm down to be able to understand." Alyssa's voice was level, and calm. It exuded assurance– it promised that they would solve the problem.

"They didn't say his name!" Steven blurted out.

She was taken aback by his abruptness. "...What?"

"I mean, for roll call." He began to pace again, "Like, in the beginning of class, they call every student's name to see if they are there? He wasn't even on the list!" Hands were thrown up in exasperation.

Alyssa stood and held her own hand out, which Steven took. They began walking towards their quarters. His mother was pondering, and Steven could see it. Everyone misses school once in a while. Kids get sick and stay home all the time. Why would this bother her son so much? Sure, this teacher not calling his name was bizarre, however it was not quite the disaster Steven was making it out to be.

"Do you think it is possible that Randall's mom called ahead and told the teacher he wouldn't be there?" She opened their door and let them inside, "That would explain why they didn't call out his name."

"No mom, I asked. I said, '*What happened to Randall?*'"

"And?" She grabbed a couple waters from the fridge and headed to sit down at the table. Her grip was loose, casual, but her son wasn't the type to press on subjects that mattered little to not at all, even if he were a child.

"The teacher told me there was no Randall. Those were her exact words. That there was no Randall– like, at all! I told her, I said, 'ma'am he has been here every day with me.' And she just told me to stop telling stories!"

"Hmm..." Alyssa considered his words carefully, but there was less than a second before he started up again.

"–and *where* are *his* stickers and gloves?" His hands were thrown into the air again before Steven threw himself down beside Alyssa.

"What?"

"Mom, they're all *gone!*" Slowly, he was raising his voice–losing his temper. With his mother. That was unheard of. Steven hadn't realized how worried he was until he laid it all out there for her to know, and even then with his hands shaking, he found it hard to believe. After all, this made no sense, right? Surely it had to be a misunderstanding.

He drank some of his water and stared with big, wide eyes, begging her to understand how alarming this was.

"Okay." Alyssa spoke after a long while. "Maybe he was moved to sick bay or something? I'll look into it. Just give me a few days," She placed her hands on his shoulders and felt them loosen. It was a little bit of a comfort, but it wasn't enough.

"Mom..."

"No transports have left or are scheduled to leave. This means that he is somewhere on the ship. Honey, I'll figure it out. Just give me some time, and don't worry about anything today."

This, more than her blanket assurance, seemed to truly calm him. His mom was taking him seriously, and she *was* the Captain. If anyone could figure it out, it was her. The transition from what had transpired to something less chaotic was slow, and steady. Together they cleaned the house, starting with the easy things–

stuff that Steven could focus on and use his hands tediously for, and stayed busy until his father came home.

When he did, it was with a boisterous call to the family. "Whew!" James threw the door open, apologizing for the way it slammed against the wall quietly, and then: "Baby! I am beat!" His bags dropped to the floor with a light thud, and he closed the door behind himself with the heel of his foot. "Give me some love!"

He wrapped his sweaty body around Alyssa and she squeaked like a mouse, which was cute considering the woman was a tiger at the helm.

"Oh! Ew! James! You smell like boy!" She yelled, trying to get out of his hold, squirming against him as he rose to the play and held her tighter.

Despite not knowing where his friend was, Steven managed a laugh. There hadn't been anything said about the discussion that was had before his father had come home, but he was still thinking about it– and the moment his laughter stopped, he remembered. Slowly Steven backed away to head towards his bedroom, trying to be inconspicuous.

Still, though, his father noticed. After all, they were still due to play ball together.

"Hey kid, where are you going?"

"Homework. Come get me when you're ready?"

"Yeah, sure." A response to his son who was well already in his room. James turned and whispered to his wife, brushing her with his fingertips. "Take a shower with me?"

The hairs on her arm rose in excitement. "Mmm. Yes, then we can talk about our days."

It was some time later, but when they opened the door to leave the shower, steam rolled out of the room.

"So. We are having... visitors?" James asked, his fingers raised to make gestures with air quotes.

"Yup, and no one seems to think that it is important that I know anything about who they are, or where they are from. We are supposed to discuss it within the next few days. Meanwhile, none of our technology problems are *apparently* a big deal. Which is actually really annoying. I can't properly do my job if they don't do theirs. And the problems," Alyssa turned to face him, tossing her hands out to either side, "*Are* a huge deal. Everything is being affected. And the longer we ignore them, the worse they get." Her body sat on the bed with a light sound against the comforter, and she reached for a nearby bottle of lotion.

"And of course, they just want you to 'handle it', right?" He did the air quotes again, which made her roll her eyes.

"Of course— and you know what else?"

"What?"

"It has me nervous. I'm not sure why, but I have a really bad feeling about it all." Alyssa didn't even realize it when her palm pressed against her cheek. "It doesn't make sense to me. No one cares that we're not running at one hundred percent even though we have these 'visitors' coming."

For a moment, she let herself pout. This was her private, safe space– she could do that here. "We're in space. The first group meant to be living out here. Our ship isn't working right. And there are strangers coming? Obviously not from Earth, so what– *aliens?* Aliens we don't *know?* Why wasn't I told about any of this sooner?"

"What can we do?" James murmured, his voice soft and understanding. It was true he was also an occupant on the ship, so this frightened him just as much if not more than the fear she had expressed– but he sympathized with the stressor that was the position she was in, too. Kneeled before her so they could be eye to eye, he took the lotion bottle from her and set it aside, taking her hand in his own.

"I really don't know." It sounded pitiful. After all, the captain should always know. *She* should know. "I guess we just keep our eyes open and work what we can."

"Like we always have." And like that, she felt less pitiful.

"Like we always have." She echoed, a soft smile playing on her lips before it stilled to a thin line. "Did Steven tell you about Randall?"

"What about him?"

"Steven thinks something has happened to him. As in, *happened* to him. Like, he's disappeared or something."

"What?" He asked sharply.

And there it was again, the anxiety. She got up and started pacing. "I guess he wasn't in school today and the teacher said no such kid even existed."

"That doesn't make sense." James murmured, incredulous, taking her place on the edge of the bed.

"I know. I just don't know what to make of it."

"You going to call Earth Point?"

"Yeah. It's all just too weird and... coincidental, I guess. You know? All of this at the same time. And where would he even be? There's no way for him to not be on the ship anymore, and we knew he was here to begin with, so claiming he wasn't—" She took a breath, frustration growing. "I want it officially documented. Even if we find him, I want it documented." There was no hesitation in her actions as she stood up to get dressed.

"Let me know?"

"Of course. Meanwhile, make sure Steven has a good evening. He doesn't need to worry about his friend just yet." Alyssa's sincerity in making sure that Steven was feeling alright made her husband smile. She was such a good mother, and that was nice to see– but with everything she'd shared, it was going to be hard to pretend it wasn't on his mind.

While James left to spend time with their son, Alyssa thought about her next move. The obvious thing to do was to call Randall's mom and see what she said. That would snuff this out fast. Why hadn't she done it sooner?

"Hello Nancy! This is Alyssa, Steven's mom?" There was a moment of silence.

"Uh, yeah. The Captain." There was more silence as Alyssa listened to the woman on the phone, waiting for more familiarity.

Alyssa felt her feet begin to walk her around the room again. "Oh no, Nancy, you didn't do anything wrong to warrant a call from me. I'm not calling as a captain right now. I'm calling as a mother." There was something strange here. They had talked before and it was never this forced. Had Nancy done something wrong...?

"I was calling to see how Randall was." The inquiry was met with silence. "Randall. Your son? He is friends with my son, Steven, as you know– and I—"

Nancy abruptly cut her off– something about how she must have had the wrong number, or misremembering something. *What?*

"No, I'm quite sure I have the right number. Your son is Randall Silvey. He goes to school with—" Another interruption, insinuating they weren't parents of any children.

"We've all had dinner together!" She snapped in frustration. It didn't feel like she was getting anywhere. Infuriatingly so, especially as this made no sense at all. "Look, I am just concerned about your son." A final silence while Nancy spoke– this one more firm, forcing her to back down.

"...Okay, so sorry about the misunderstanding. Have a good evening."

Alyssa hung up the phone. *"What the fuck?"* Cursed under her breath, finger darting up to the handwritten note bearing the Silveys' phone number. She hadn't been wrong.

Almost at the perfectly wrong time, Steven came into the room, and Alyssa quickly tried to fix her scowl into something else– something he needed right now, not this.

"Mom, dad is putting on his shoes. Will you look over my homework while I put my shoes on too?"

"Yeah, sure, hun." As if nothing was wrong, she smiled at him. Looking over the questions was meant to be a quick task, but the words were blurry to her.

"You look like you're about to start a fire in your mind." James noted as he came in.

"Oh." Alyssa's eyes darted to James. "Can you look at this? I just can't seem to focus right now." The paper was placed onto the bed while she stood up to pace again.

"Hey." Arms were put around her in a comforting gesture. "What happened?"

"I don't even know... but I'm beginning to think Steven is right."

CHAPTER TWELVE

All that Justin could think about when he got to his new home was that whoever had the contract to build the houses had an easy payday. They were all the same cookie cutter shape and size. He was given his key and a look by Alex.

"So uh– there is plenty of alcohol inside. You and Jeff are on mids tomorrow... so you don't have to come in until three." There was a forced grin on his friend's face. Out of the three of them, Alex was the only one who didn't drink. If he was the one who had bought the alcohol, they were all in trouble.

There was nothing left for Alex to say or do, and Justin felt a little awkward with standing and continuing the conversation, so he lifted his hand in a wave and walked to the door. Before he opened it, however, he stopped and took a deep breath.

Now that Alex had already gone, it gave Justin a moment to collect himself. Once he started tomorrow, he would be watched like a hawk. But, for tonight at least, it seemed as though he had a reprieve.

The man walked inside and looked around. He hated it here already, filled with that feeling of not wanting to be there. Justin placed the key on the table and looked at the line of liquor that had been left out for him. Immediately after, he went over to the fridge and opened it up. Oh look, *more alcohol.* This one had a variety of beers and coolers, with a couple of wine bottles. No doubt Alex was trying to make sure he was stocked for a while. He grabbed a six pack of beer, not even noticing the kind it was, and went over to the liquor. Jack Daniels would do just fine.

There was no effort in locking his house as he left for Jeff's. His pace was slow as he walked. He wasn't sure he would be welcome. Alex thought he would be, but he wasn't a part of the last encounter they had. Justin thought back and shuddered.

Jeff grumbled as he came into the jail hallway, "What the fuck, Justin?"

"Nice to see you too, *friend*." He emphasized the last word sarcastically.

"Fuck you." The man growled at him. "You did so much damage that I'm feeling it in my ass half way around the fucking world!"

"Did you come all this way to yell at me?" Justin lulled mildly. He was sitting on the floor, even though there was a bed at the end of the cell.

The room was quiet. The other man didn't bother to apologize. Justin knew it by his look. He could never stay mad. He even knew why he'd come. In truth, his 'friend'– who truly was a friend– looked tired, hungry, and scared.

"Damnit." Justin muttered. "Look. I didn't mean for any of this."

"Oh. Right. You just meant to tell them what a happy family they would make on the ship?" Jeff rolled his eyes. He finally made eye contact with Justin, though, which was a step up.

He could get a proper look at his friend this way, at least, something better than a poor glance. But he deliberated on if it were better he did or not. Yes, he had gotten the idea of his state from afar, just focused now... he couldn't get over how jail made him so... *haggard*. Justin was usually pale and gaunt, but, not like *this*. Never this bad.

Jeff asked quietly, "How are you?"

"I'm..." Justin paused, thinking of what words to say. "...Okay."

"Don't do that."

"What?" It wasn't a question

"You look like shit. You smell even worse. If I didn't know any better, I would say you had been–"

"What?" Justin retorted quickly, "Tortured?"

"Were you?"

"No," He laughed softly. "*No. That might have been easier to handle.*"

"Then..." Jeff inquired more than he'd meant to– asked without wanting to actually ask.

Justin mused while looking off, tapping his fingers idly on the floor. "Really– it hasn't been terrible. I've been in intense therapy since I got here."

"Intense therapy." Jeff repeated.

"There isn't much to say. Every day I have to talk. To listen. To re-think my thoughts and change how I see things."

"How you see things?"

155

"Are you a parrot now?"

"No. Sorry, just taking it all in."

"Hmmm..."

"Go on." He said, finding a place and sitting crossed legged on the floor.

"I want to know more of what you think and what has been going on instead of telling you more of my end." Justin said while continuously tapping his fingers in the same manner.

Jeff stared at him dead in the eyes. "I need to know you're okay."

"Why?" Justin asked. There was something prickling at the base of his neck. Jeff was rooting around for something.

"You *are* my friend." Jeff replied, a bit taken aback.

"No, really, why are you here?"

"What?"

"Who sent you? Or what are you after?" Justin got up and started pacing. There was definitely something wrong about this. He wasn't crazy. Jeff was here for a reason. There was no other explanation.

"No one sent me, Justin..."

"Then why are you here? You hate people, yet you braved the crowds to get on a plane? You hate prisons and jails, but here you are sitting on my floor. Something doesn't make sense, *friend*." The last word was almost accusatory.

There was a pause, and instantly Justin knew he was right. The other man sighed audibly– he looked resigned.

"I need to ask you something." Jeff finally began after a pause. In truth, he had been debating even asking at all. Justin didn't look too good, and he was scared to throw him over the edge.

"Ask." Justin demanded, looking almost... deadly as he did.

"Will– ah," He took a breath, then spoke so fast that it all fell apart into nonsense; he wasn't anyone's first choice for auctioneer, this was nonsense. It had to be repeated a second time, which Jeff finally understood: *"Will things work on the launch? Are the numbers really correct?"*

Slowly, Justin's lips curled into an expression Jeff had only seen in the movies. And just like them, Justin began laughing… maniacally. Though cut between the harsh noise was a deeper cruelty– curses, unhindered snarls of them, each scathing word he could get at Jeff until he finally left.

Then Justin found himself alone again, on the floor, in his jail cell.

Justin came back to the present when he saw a figure walking towards him through the thin panels of the door windows. He'd managed to get up the steps and knock on the door, but he wasn't sure when he'd walked the full distance.

"Yeah?" Jeff called out without opening the door.

Justin hesitated. What was he going to say? He didn't think this through. How was he supposed to break the ice?

"Uh– hi. It's me, Justin."

The door unlocked and slowly opened to reveal a shocked face on the other side.

"Surprise." Justin spoke sarcastically and cautiously.

"I. What– Um... Why?" Jeff took a breath and rubbed his hand over his face and seemed to snap out of the shock. "Oh hell. Come here." There was no more time for hesitation– feelings aside, it was time for a hug. "Come inside! How are you? When did you get back?" Jeff started the 'catch up' while still patting his back in their embrace.

"Oh. I got back today. And... I'm okay, all things considered." Justin stepped back and rubbed his neck.

They were both awkward people anyway, but their social skills were so deplorable that they ended up staring awkwardly at each other, away, and then back again.

"I brought alcohol!" Justin announced while raising his hand to show the boon, his mouth lopsided in a sheepish grin.

They both went inside, put the Jack on the table and then just looked at each other. It was silent for a long while, the space filled with the tangible, shared thought: where to even start?

"Welp. I need a smoke." Jeff announced, closing the door behind him.

"Me too, actually." Justin thought aloud, both leaping at the chance to do something with his hands, and reacting wholly out of habit– though it earned him a sharp look from Jeff. "Yeah, sue me. I think now we officially violate some world law. Smoking is so taboo now, you notice?"

It was a ramble, evidently due to fear of judgment. Sure, the man before him smoked before, but that had never been Justin's thing. The two had shared jests over it in the past. Justin hadn't thought much of it until now, but he was afraid of what the other would think.

Jeff rose a brow, but with a sly and curious smirk. The reality was, that they were both nervous– and that the ice was technically broken. They were just easing into the friendship that they'd had previously. *Friends*, after everything.

The two walked through the house, which Justin noticed was an exact replication of his own, even all the way to the back door. Outside there was a patio table with chairs, and an old coffee can that Jeff clearly used as an ashtray.

"So, did you go to Earth Point yet?" Jeff could help but ask while handing Justin a cigarette.

"Sure did." Justin answered, fetching the lighter off the table.

Their catching up ensued, with both talking about the punishments that were given to either of them by One World. How much they both hated Clinton was also a big portion of the conversation topic. They went through two beers a piece throughout the whole thing, then were silent for a while. The best thing about their friendship was that they didn't always have to talk.

"Are you moved in?" It was inquired tentatively.

"Yup."

"And you, uh—" Jeff hesitated, the air turning tense, "...brought everything?"

Justin gave him a look, as if worried the walls could hear, then nodded once. They were silent again for a time. Both knew that the unasked question was if Jeff had brought what had been left behind the furnace. One drunken night, Justin had given it to Jeff and told him to keep it a secret, and to absolutely keep it safe. Jeff never asked any questions about it, and they never spoke of it thereafter.

CHAPTER THIRTEEN

"Today is the day, guys." Alyssa's voice filled the room where her team had gathered.

There was an eerie silence to the group. As the day approached, more and more things seemed to break. First it was the connection with Earth Point, and then the bay doors. Then it was the heating and cooling systems. The water rose to such temperatures that the pool hall turned into a sauna, and then everyone had fallen ill all at once with the constant cold air.

But no matter what happened, the visitors were determined to stay on course to their ship. Alyssa was just as nervous as her crew. She couldn't show it, though. Not now. She had to have all the answers... know what was going on– who they were. Or, at least for her crew, she had to fake it.

"Mary, what time is it?" Alyssa looked over at her second in command and offered a smile.

"Zero eight hundred, Captain."

"Okay– hey," She paused, looking around the unit. "Anyone know where Miles is?"

"I went by his lodgings before work and the whole place was empty. I figured he finally got his room with a view, or that he

161

moved in with his girl, finally– so we could have more room for the visitors." One of the crewmates, Joe, answered.

"No one has had permission to change rooms in nearly six months."

What the hell? Alyssa leaned back in her chair and tried to hide her surge of anxiety. Immediately, she knew that this was no coincidence. This was now the second person to go missing on the ship, including Randall– who they never found, not to mention the lack of evidence that he was ever there to begin with. She stopped her thoughts before they could run too far. "Joe, do me a favor and go look for him. Start with the girlfriend. If you don't find him on your first sweep through, come back. I need all hands on deck." There would be no flying off the handle yet. He could just be doing... something. And Alyssa needed actual confirmation first.

"Direct, Captain."

"We don't have a lot of time before our visitors arrive. Where are we at with our rebuild?" Alyssa had been asking this question every day for the last few months.

The list that was given to her of all the things that were wrong was absolutely ridiculous. For every problem they solved, there were new issues that popped up. At least, the bay doors had been fixed. They wouldn't have a problem with the visitors coming in... but their contact with Earth Point had been spotty at best, and never held long enough for a full brief by either party. Their electronic correspondence had been a little better, though.

Alyssa went to her office and pulled up her last transmission from Earth Point.

"Captain. We are working on the video feed, but as of right now, all systems are a go for you to receive the visitors. Do not disappoint us."

The message bothered her greatly. *"Do not disappoint us"*. Like she would ruin the 'family dinner' or something? Nothing was said at all about her concern with Randall, and as the communication was able to be sent at all, he could have written as much as he wanted.

Alyssa ended the recording and sat back in her chair. There wasn't time for her to be idle, though. She was making mental notes in her head. Soon the list became too long. Grabbing a notepad and pen, she began to write it all down.

After having written two pages worth, she called Mary in and passed it to her. There were larger tasks that only she could do, like attempting to contact Earth Point on her encrypted tablet– and there were smaller things that other people could handle, like double checking the status of the visitors' quarters. She had no problem delegating work.

Time passed, feeling faster than she'd hoped. Finally, the time had come for the visitors. Joe had come back and reported that he could not find Miles anywhere... and his girlfriend acted like she didn't know who Miles was. Just like Randall's mom had. But before Alyssa could get into that – and she really wanted to – they had picked up the alien ship on radar.

"Put me on main comm." Ordered aloud from the control chair. She knew that it would be followed immediately.

They would be landing in the bay in less than twenty minutes.

"This is Captain Burns." Her voice resonated across the entire ship. "As you know, our visitors will be here soon. I know this is exciting and strange. No one has ever met a being from another world." A pause. But she could not afford to hesitate, for her own reservations– "There are a lot of emotions that come with this. But we were chosen. We were chosen for this mission, and chosen to meet these beings."

Alyssa took a breath, steeling herself. The clock ticked on. They were an ever-nearing entity she knew nothing of, and could do nothing to halt the arrival of.

Her orders were as spoken. The captain was to follow them: "So, now it is time to live up to what we were chosen for. I need everyone to the bay's viewing hall. Our guest will be docking soon. Please be quick and safe. Let's give them a warm welcome! Burns out."

People all around were bustling and getting ready to go to the designated meeting place. Joe came up to her with multiple tablets in hand, with one being given to Mary and the other one to her. They unlocked their respective tablets with their fingerprints, ready to go. With these devices, they'd be able to run the entire ship.

On the way there, Alyssa led at the front, the crew following in her stead. Joe was looking at his tablet, making notes to Mary, and swiping away with his stylus. The captain knew that her people had things covered, so she only caught a few tidbits of their conversation– she didn't have to listen to it all. Soon, it became mingled with the voices of other crew members.

"Twelve point two clicks."

"Radar active."

"I wonder what they will look like...?"

"Mom, do you think they eat apples?"

Alyssa felt herself smile at the general positivity of the ship. There had been worries about how everyone would react to the news of aliens coming. There was so much that she had been worried about for so long because of this visit. But it was finally cresting over. The hurdle would be jumped, and all would be well. At least, she hoped.

Everyone had their orders to be in the side viewing hall, and there was a moment where she could let herself feel pride to see how well everyone had gotten to their respective spots in the room, like they'd been told. Only a small crew of about ten would stay with her, their eyes and fingers working away at the screens in their hands– letting her focus on the alien visitors arriving.

She could see the ship docking through the windows of the entrance bay, doors closing behind it. A jet of smoke blew from its back end, and filled the space around it, concealing itself from everyone's view. The smoke turned into wisps as it slowly dissipated.

And then she caught sight of them.

One by one, their bodies pushed through the concealant. They were tall. Very tall. Each had long, deep colored robes. The color was beautiful; almost like mahogany. It was the most gorgeous blend of burgundy and brown she had ever seen, both majestic and earth-toned... they did not have textiles like this at home, none that she had ever seen. Still, this lavishness had an 'off' detail: she could not make out their faces with their hoods up. But, she figured she would see them soon enough.

"Joe, open the entrance doors." She exhaled.

Here we go.

The doors opened with a whir, and Alyssa stepped forward to greet them. Throughout her career, she had been trained to meet all sorts of people. She knew how to address those that used to be royalty, dignitaries, and just about all others. But she had never been trained on the customary greetings of... otherworldly beings. She thought back to her training, though, and went with some very basic suggestions that had been given for foreign leaders. Thinking about it now, in this brief moment, she realized it was strange she'd never been given training for meeting aliens– especially now that they were here, and Earth Point seemed informed of their arrival.

"Welcome to Space Port." She opened her hands in a greeting, and added a smile before bowing her head slightly.

"Thank you. I am pleased to be here. These... are all of my house." The voice did not so much speak as it did echo.

Instantly, Alyssa stepped back– almost recoiled from the greeting altogether. The woman's warm welcome faltered. She couldn't place her finger on it, but something was... wrong. Visceral– all of her nerves were screaming that she was in danger. *Danger.*

Thankfully, it was only a split second of thought and reaction, and no one noticed. She recovered by turning while she moved back, and motioned to her crew, smile as-promptly returned in full grace.

"My crew and all aboard the ship welcome you." It was her attempt to sound gracious and without suspicion.

"There is much to see and do, but I wonder– if I may ask to have some time to recharge before commencing? It has been a long flight." The stranger reverberated, slow and steady.

"Oh, of– of course." Alyssa stammered, caught off guard. "We were about to break for dinner. Do you want us to wait for you? Or perhaps take the food to your quarters?"

The echo snapped, "We do not require sustenance."

If Alyssa didn't know any better, she would say that the being was annoyed with her asking.

"Of course."

"You will show us to our rooms." They thrummed– sickly sweet, and then added, *"Please."* in that same tone of false sincerity.

"Of course." Alyssa repeated her previous atonement, and then turned to Mary. "You have the ship."

"Direct, Captain." Mary turned to leave, and Alyssa wished she could have her people stay.

She felt like she was in a den of snakes. It didn't feel right.

"If you would, your quarters are this way. Do you prefer to lead, or would you like to walk beside me?" She figured it was the best way to not offend whatever their custom might be.

"You will lead. We will walk behind you."

The other beings behind the speaker began to communicate. However, it wasn't with words. It was a cross between a snake's hiss and the *shhh*'ing sound of a hush. The being that had been talking made a click-like noise, something like a direction or command, and abruptly all the other beings stopped. The air was awkward. It felt like being at a friend's house when the mom of your friend was yelling.

Discomforted, she tried to break the silence, "We will go through the entire ship during our tour, whenever you are ready. But, coming up are the living quarters. The first hallway is yours."

"The entire hallway?" The apparent leader inquired. "I do hope we did not put you in any trouble to give us all of this space."

167

"Umm, no. We have extra lodgings available, so it really was not an issue." The hairs on the back of her neck stood up. She didn't like having them behind her, where she couldn't see.

"Wonderful."

From then, they moved in silence. Noticeable, even if so brief... the hallway came up quickly.

"Well, here we are. You can choose to room in twos or all separately. All of the doors are currently unlocked, though the keys are on the counter, should you wish to change that."

"My thanks to you."

"I'll let you get settled. There are intercoms on the walls of every room. When you are ready, please push the blue button and it will link to the command room. That is where I will be."

The being slightly bowed, and when it did, she caught a glimpse under the hood. It was a mask. No. Not really a mask, but a solid piece of white that covered the entirety of whatever was under the cloth. There were two slits cut horizontally across the plank. She suspected that this would be for the eyes... and perhaps the nose?

Alyssa dipped her head in a bow and turned to leave. It took everything she had in her to not run. She was nevertheless brisk, *ah*, in the way sportswomen were she supposed. Leaving this and entering the next, she came across a maintenance worker in the hall – stopping as the realization hit, like putting a few pieces together – it was the one and the same that her husband had previously mentioned.

"Mario!" Alyssa exclaimed. She was so happy to not be alone so close to those beings.

He nodded in acknowledgement. "Captain."

"Hey, look–" A glance towards where she left them. "The visitors are in the first hallway. Don't go down there."

"...Captain?"

"I just don't want to disturb them, if we can help it. They won't be here long, hopefully."

"Ah, okay. Just let me know if they end up needing something. I would be happy to–"

"No. Don't go there."

"Captain, are you okay?" Mario's expression seemed to be searching hers as he reached out to put a hand on her shoulder.

Shit. There must have been a certain terror in her eyes, a fearfulness that could only convey 'threat'. She'd... more than once been told she was obvious about it.

"I really don't know."

"I am here if you need anything."

"Thanks Mario. I just... need to get back to the control room."

He nodded and left in the opposite direction of the hallway.

Alyssa took a few self soothing breaths before she cut across the ship to get back to the command room. She was churning the experience over and over in her mind, but when she opened the command doors, it was to applause and people smiling and congratulating her.

"Great job, guys." She forced a smile, trying to appear as happy as she could. "You guys take a bit of a break. Go eat. Mary, Joe and I will follow behind shortly."

Everyone filed out while Alyssa went into her office, followed by Mary and Joe.

"Yes, you have permission to speak freely. I need to know what you guys think about our guests." Without having to be asked, the captain immediately jumped right in to inquire her closest crew's thoughts.

"That was... odd."

"They gave me the heebie-jeebies."

"How did you know they were wanting to go to their quarters?"

"I was wondering the same thing."

"What do you mean?" Alyssa was confused. "They said they wanted to go to their rooms. Didn't you hear them?"

They both replied in unison: "No."

"They were talking to us normally– albeit with a weird, echoey voice. You didn't hear?"

"All we could hear was a *shhh*. It was like a whisper." Joe commented.

"Or a snake?" Mary countered.

"The echo."

"What?" Joe asked.

"That's why it sounded like an echo."

Alyssa couldn't get another word in– alarms were sounding, one beep after the next. It wasn't a red alert, though it did get their attention. Joe walked over to check the screens.

"Captain, we have a problem."

Mary was typing on her computer next to the other man when Alyssa walked up, looking over her shoulder. It was then that Mary turned to her. A chill ran down her spine, this was not the expression she was accustomed to, not from Mary. It was not that urgent, not of what she'd heard, then what–

A breathless statement: "Earth point has just gone dark."

CHAPTER FOURTEEN

"I need coffee." Jeff mumbled as he stared into an empty mug.

He had been at the console for a little over twelve hours now. Couldn't sleep— hadn't been able to since Justin came back into the picture. The truth was that he was glad to see his friend, he really was. But people kept treating them both differently now that he was back. On one hand, he didn't really care. Jeff had never been close to any of the people here. Though, on the other hand, it wasn't like they were just leaving him alone as they normally did. They were avoiding him... which was new. And, in Jeff's experience, anything 'new' was a bad thing.

He stood up and stretched so wide his back cracked. Letting out a yawn, he grabbed his mug and made his way to the break room. He was too tired to think rationally– especially about something that could lie. People weren't just mere numbers. There were more variables with them, and now there were too many things that kept spinning and spinning in his head. There was no way to process them, though. There were too many things wrong for that. Not enough sleep, not enough work done. All he could do for the moment was get coffee– surely he could do that.

Except, he never made it to the coffee pot. Alex was coming towards him, looking grim and grumpy.

"Who pissed in your cheerios this morning?" Jeff spoke sourly as he tried to walk around him to get to his drink.

Ugh. It was empty. With the pot in hand, Jeff made his way to the sink area to rinse it out and add water to the machine.

"Jeff. I– we need to talk." Alex's voice cracked as he spoke.

"Who died?" A sarcastic question from Jeff in return.

It had been incredibly noticeable that Alex had been really glum lately. The man suspected for a while now that the other knew things that he wasn't sharing. Whatever it was, it was making him as jolly as the grinch. But, when he turned and truly saw Alex's face today, right now, he knew something was... it wasn't just a bad mood, he could tell that much. Something happened.

"Oh god– who actually died?" Jeff nearly dropped the coffee pot in an attempt to place it down, his heart thrumming.

"We need to go somewhere more private." Alex grabbed his arm and began leading him to a nearby office.

Where was all the air in this place?

Jeff didn't want to know who it was... though he was pretty certain he already did.

"When?" He breathed, looking at Alex.

"I'm not sure. There was a call about a disturbance. MP's went to his house."

Jeff sat down on what he thought was a chair. For all he knew, it could have been the floor or a table. "*How?*"

"I don't know. I don't have the specifics. But I am giving you the day off. Effective immediately. And tomorrow." Alex shuffled his feet as he spoke. Jeff felt as though he looked like he just needed something to do. It was the only way he knew how to help.

"I need to go to my station and grab my things." Jeff murmured. He didn't get up from where he sat.

Alex moved to sit beside him, which appeared to be a deliberate motion, "We can just sit here a minute."

Jeff realized, then, that he was in fact sitting on the floor. His friend sighed and looked around his knees at his shoes.

After a moment of silence, Jeff finally spoke: "I need details." However, he hesitated to say more for a long moment. Still, "I know it's morbid. I know you may not be able to give me all of them– but I need to know. I have to finish the story in my head. Otherwise, my mind will create its own scenarios."

"I'll see what I can do. But for now, just–" Alex breathed deeply, "I don't know. Sleep. Process it and then come back to work. I need you here."

Jeff snorted. "Yeah. I'm really going to be able to sleep."

"Then read a fucking book. The bottom line is that you are not going to be effective here right now. So let me take you home. Please."

"Yeah. Okay. Help me up." The more that Jeff allowed himself to think, the more he felt sick to his stomach.

To be honest, Jeff didn't really register going to his station. He couldn't tell what he did at all. Did he turn off his computer? Did he close his notebooks? Did he cap his pen? No idea. He did know that Alex let him smoke in the truck though. And that meant he looked as bad as he felt. He was never allowed to smoke in Alex's vehicle.

When they pulled up to Jeff's house he didn't even realize it until Alex let out a cough. He thanked him and got out of the truck, but before he closed the door, Alex raised a hand to stop him.

"Hey Jeff, do me a favor?"

"Sure, Alex. What can I do you for?"

"Don't go to Justin's house. I don't know what happened and I don't want you walking into a mess."

"I uh— I hadn't thought of that. I won't. Thank you." Jeff realized he'd left his jacket, grabbed it quickly, and closed the door a little harder than he meant to.

As he headed towards his home, he didn't quite want to have Alex wait for him to go inside. Every part of himself felt that impulse urge to look at Justin's house; take his time to think about it all. So, Jeff made a show of sitting on his steps and making himself busy, pulling out a pack so that he had something to do with his hands. Honestly, he didn't really care to have another one, but he knew it would get Alex moving.

Soon enough, after realizing that his friend had chosen to instead linger outside rather than head inside immediately, Alex raised a hand in a wave, and drove off down the way.

Jeff looked towards the house that rested besides his. It was his chance now. And, even with that, did he really want to pay that place a 'visit'...?

He knew he had to. Probably before the morning light broke, because chances were, people would be there in the morning. When it came to these houses, they were never empty for more than a day or two. No, it would have to be tonight that he investigated.

Yet... while it may have been paranoia, he felt as though he were being watched. Piercing eyes on his back— he couldn't shake the sensation. But Jeff didn't know why he thought so. Alex left about a minute ago, he was *sure* of it... and it didn't matter that he had somehow, not at all. Might've been worse than when he *was* there. He could rationalize it a thousand times and still *feel* it. Something told him he had to keep up his act from earlier– to make a show of going inside and going to bed. It was scrutinizing him, and very carefully so. One wrong move and... he wouldn't think about the consequences. They felt too real.

So that was what he was going to do: what he was instructed to.

He was never the stealthy and cool type. Jeff was going to have to make do, and there wasn't a choice if he was going to make it into Justin's right now.

He wondered why Alex didn't want him going over to that house– had it not been cleaned up? What was the secrecy for? The fact that Alex didn't want him there made it impossible not to go, like he was compelled to. But– that feeling reminded him he just was not *supposed* to, and that had nothing to do with Alex, did it?

Jeff smashed out his hardly smoked cigarette and went inside, turning on the kitchen light. He realized that not once had he been inside Justin's house; he didn't know the layout, though chances were it would be the same as his own house, thankfully. They were all made the same way, dozens of popups... That didn't stop the nervous feeling sinking in his chest. He went and grabbed one of the open bottles of Jack Daniels and took a swig. Ah, well– no, he wouldn't do any more than that. He needed to stay sharp.

Feet took him upstairs where he took a shower, and then turned off all of his lights except for the one by his bed. The soft glow was enough to read by, but not enough for much else.

Which was perfect. He didn't need anyone to be able to see that he was, in fact, not sleeping. He laid in bed with the book open for a few hours. Idly, he flipped through the pages.

Then, he decided, it was time.

Crawling out of bed slowly, he made sure to stay low to the ground. Bear walking towards the door, he was thankful he remembered to keep it mostly open for his escape. He had stayed fully dressed, so no need to put anything on. Once he reached the hallway, the man rose to his feet. There were no windows in this area so Jeff was able to stand up-right and walk as usual. Though, he was still having trouble breathing properly. What would happen if someone was in front of his house? Or in Justin's house?

Once in the kitchen he hid in the shadows and passively wondered if he may need a weapon. He didn't want a knife. There was nothing that said *'intent to harm'* like a knife. So he grabbed what he thought was the next best thing, and went for a rolling pin. He grabbed the handle of his door and slowly opened it. As silently as possible, his hand grasped the screen door and pushed it open. The slide was new enough that it didn't make any creaking noise, but the pounding of his heart would have thought otherwise.

Outside, he almost choked–it was too heavy, his breath caught on the first step. There were eyes all round, he could just feel it. Or maybe it was the lack of sleep and the idea of what was about to happen. *Shit*–faster than fear could stop him, he hinged on logic: Jeff and Justin's gates were always open, and it was in such a way that it was totally encased in shadow. That would be helpful to block out any unwanted observers. The sigh of relief Jeff let out was a little too audible, but he continued anyway. He had successfully comforted himself– given himself the hype needed to move forward. Hands shaking were to be overlooked.

Finally, after what seemed like forever due to the slow nature of his steps, he caught sight of the back door. He was in his friend's yard. Jeff approached cautiously, taking a quick glance around and noticing nothing. It was silent. Moving towards the back entrance, he noticed the folding chair and tin can that had been used as an ashtray. Feet forced onward, he extended his hand towards the door and stopped.

What if there was blood? What if they hadn't moved the body yet?

Nausea started to bubble up from within him. Dizzy. He couldn't do this. Trembling legs sat themselves down on the folding chair and Jeff quickly grabbed one of the smokes from a pack that had been left on the floor. He grabbed the lighter and inhaled like his life depended on it. There were no coherent thoughts going through his mind, but by the end of the cigarette he felt a little more 'with it'.

No matter what, he was getting into that house tonight.

Looking up at the sky, there was a slight breeze that brushed through his hair. That was... in a weird way, all the push Jeff needed. To him, it was the closest thing to a nod from Justin that he would ever get; a strange notion, considering he wasn't a spiritual man, but he didn't dwell on the feeling. He inhaled– deep this time, and pulled the outdoor screen door handle. It clicked once, and when he turned the knob, it fully opened without a sound. Didn't think it would be locked. Justin was never the type to, anyways.

Jeff couldn't turn on any lights, because as it felt outside, he knew that this house was more than likely being watched as well.

This was both a good thing and a bad thing. Good thing, because he didn't have to see if there was a mess from whatever Justin had done to end his own life, and... a bad thing because he wouldn't be able to see if he stepped in anything *liquid*. Crouching down, he decided to try and use the streetlights filtering through the windows and ambient glows of the night itself to see his way around. Making out the tile, he felt like it was safe enough to continue without the fear of stepping into something that would leave a trail.

Though as he looked through every room of the house, he became gradually frustrated. There was nothing to be found. His heart thrummed out of control every time he had to open a door, thinking there would be something hideously gruesome on the other side, only to find it... *empty*. The bathroom was the only thing that had any sort of mess in it, and it wasn't the death kind of mess. Just the mess of a man who didn't clean up after himself.

Now that the initial search was over, however, he could go through the house again and see all the things that might be out of place.

While there had been no viscera anywhere, there were a few immediate red flags. Justin's computer was gone. His notebooks as well. Books were left behind, though, and that is where Jeff went to investigate further. Peering at the names, he searched until he found the one he was looking for. 'Great Expectations' was a boring read for Jeff, but he and Justin had used it for years to leave notes to each other so that no one else would find them.

Opening it up, he found a torn piece of paper with a hastily written note–

"Don't believe what you're told. Keep it safe."

Jeff pocketed the book then and there and left. He was just as jumpy leaving the house. Keeping to the shadows, he managed to get back into his own without any incident and quickly decided to actually sleep this time. He did his job, or so he swore he did. He didn't think he could handle more of it, fear exhausted him that night so much more than he thought possible, even after everything.

Before he settled in, though, he put the book somewhere inconspicuous– he had to. It was on the shelf with the rest of his personal books. It looked as though it belonged there, and no one would be one the wiser to think of it as anything special compared to the rest, let alone open that one in particular.

CHAPTER FIFTEEN

Earlier, Alyssa had told the visitors to let her know when they were ready for her, and they had yet to do so. She figured that maybe they might just be tired and in need of a good night's sleep, but that didn't stop her from feeling on edge. The Captain didn't like them. None of her crew did.

"We still can't get anywhere with Earth Point." Joe had informed her. He looked frustrated and exhausted, possibly as much as she did too.

"Okay." Alyssa replied, looking up to the roof with her hands on her hips. They couldn't do this indefinitely. A new plan of action had to be made. Something.

"Alright." She started again, "Here is what we are going to do. Joe, you pass-down what you've done, or next steps, or whatever you need the second stringer to do. Get some sleep. We can work in four hour intervals. Mary, you also do pass-down and tell your second to push through any contact with the visitors to me. I'll be taking a comm with me." Her pacing took her over to a case and grabbed a more advanced looking radio receiver. "I'll also be going to bed. I don't know when I'll be showing the visitors around but the last thing I want is to be too tired to think when

I'm with them." There was a glance around to see if everyone understood where she was at, and what she was needing.

The entire crew looked at her and nodded, then immediately got to work on pass-down.

Pass-down consisted of three parts. One was writing down where they last left off in case computers failed, typing into their log where they were currently, as well as a verbal relay of events. This helped for a lot of reasons. Sometimes, words could be spoken that weren't written down, and those are the little gems that could help solve a problem. Two, there is a running log of everything that was done, and three, there is a paper trail that can't be digitally messed with.

Alyssa waited until all of her staff had been properly given a pass-down. She let them leave to continue their tasks, then walked to each station to ensure that the second crew had everything needed. When that was done, she then grabbed her radio and departed from the room, allowing them their focus.

Walking down the ship in the middle of the night had never struck Alyssa as creepy before, but walking in the darkened halls of tonight made her feel like she was in a horror movie. Everything was quiet, and remarkably still. Everything might have seemed calm and okay, but there was an eerie feeling in the air. Maybe it was how the visitors made her feel, how it felt as though she was being studied by them. Any moment now there was going to be a jump scare at her expense and she was going to be forced to run across the hallway away from Michael Meyers himself.

There was a saying that Alyssa thought of. "Many fears are brought on by loneliness and fatigue." She was definitely alone, and completely exhausted. No wonder she was filled with dread.

Her breathing remained long and deep, like she had been taught in combat training. Her mind had to be focused on where she was heading, that's the priority for now. Two lefts and a right, that's what she needed.

There was something behind her. The realization came to her suddenly and made her stop in her tracks. Alyssa straightened, letting her senses roam until she felt the location of the onlooker. Directly behind her and to the left. In the shadows. Had she turned around, she wouldn't be able to see them. Instead, she had knelt down to tie her shoes, gaze directed to the very corners of her eyes.

The presence was definitely moving closer to her. Moment by moment, she could feel it coming up behind her. Hands rested above her already tied laces, bracing herself. Just a little closer.

"Captain!" A voice sounded in the night. It was the maintenance man.

"Oh... Mario, was it?" Alyssa's voice was strained, and she was sweeping the space behind her with her eyes even though she knew the presence was gone.

"Yes, ma'am." He bent down to her level and held her arms, "What are you doing here by yourself? It's not safe." He raised up with her and glanced at her back.

"Did you see it then?"

There was a shake of the head. "No. But I felt it. It was behind you, whatever it was." He paused. "Strange things have been happening, Captain. Since those visitors arrived, I've felt... all wrong."

"Why are you out so late, Mario?" She asked with the tilt of a head, trying to not be suspicious, but the feeling got the better of her.

"I do a lot of my cleaning now. I was just finishing up in the public areas and was headed to clock out when I saw you."

"But you didn't see what was behind me?"

"No ma'am. But like I said, I could feel it. I have been feeling a lot of weird things since they came. I haven't been alone since. I know that sounds weird, but it is what it is."

They had only been here a few hours, but her crew were feeling it as bad as she had the moment she met them. "Let's get you clocked out and we can head home together." She spoke, trying to sound assuring.

Mario looked as though there were things he wanted to say, but didn't. Instead, he fell into step beside her, silent until they came upon a computerized kiosk. He clocked out and they turned to head to the living quarters.

"We should go the long way." Alyssa requested at the same time that Mario had said, "We shouldn't go this way." Despite their tense circumstances, they smiled at each other. Both began to walk, knowing exactly where they had wanted to go. It brought them up behind the school and towards the back of the living quarters.

"Where do you live?" Alyssa asked.

"It doesn't matter. We are safe here. Let me walk you to your door." He smiled at her.

"What do you mean we are safe here?"

"Do you not feel it, Captain?"

She stopped for a moment to consider it. He was right– the sensation of feeling watched was gone. No such gaze could be felt on her. There was nothing but the ambience of the ship that she had been accustomed to.

"Thank you." Her voice was soft as she spoke, the gratitude apparent in her tone. Walking up towards her homestead, she turned to offer a small wave. "This is me."

"Good. Get inside and I'll see you tomorrow." Mario turned to leave.

"Mario?"

"Yes, Captain?"

"There was something you wanted to tell me but you didn't. Tell me now." Her eyes were steady as she looked over to the maintenance man.

"I don't know how." He admitted.

"Just– say it out loud and see if it makes sense."

"I think the visitors are doing something to the crew." Mario blurted, so fast that he had to say it again but slower so that she could process it.

Alyssa paused in her steps and looked intent. "What... do you mean?"

"I don't know yet. Haven't figured it out. But I can tell you that people are not acting normal, and I haven't been able to find my worker, Austin, since they showed up."

"He's *missing?*" She asked fervently.

"I don't know. It's only been a few hours. But I can't find him as of now and that– that's very odd."

"When did he go missing?"

Mario shook his head. He didn't seem sure, and simply said, "We will see if he turns up tomorrow, and... if not, I will put in a full report. For now, get some sleep. You look like you will fall over soon, and the halls won't be calm for long." All he could give in return was a little bow before walking away.

185

Turning to her home, she entered without really thinking of anything besides the news that Mario had shared. *Another missing person.* Were there any connections between them...? As the door was closed, the woman double checked to ensure that it was locked as well. She added the chain tonight. They had never bothered with the privacy chain before, but it seemed to be necessary now.

"Alyssa?" James called out to her quietly from the galley kitchen.

"Hey baby," She responded, feeling the relief hit her instantly by having him close to her.

When he spoke, it was slow as he eyed her cautiously. "How did it go?" He knew something was wrong.

"Umm... to be honest, I don't know. I don't quite like our guests..." Alyssa basically slumped into her husband's arms, feeling herself decompress.

"Oh? We couldn't really see them. And when they didn't have dinner with everyone, well, I figured that they were probably tired from their long journey."

"Ah, they did say they needed some time, but there's definitely something off with them. They put me on edge." Alyssa walked around the bar area of the galley kitchen as she spoke. Grabbing herself a glass of water, she soon returned to sit at their little table together. "I was on high alert with them, James. Like I was worried they would very literally stab me in the back. I don't like them here. The sooner they leave, the better."

"Do you think it is a real 'high alert' or just a new species that you know nothing about?" James asked and sat down next to his wife.

186

"I'm not sure I know what you mean?" She turned to look at him.

"Like a snake? Not all of them are deadly. But, when you or I see them, our first response is to grab our gun because they put us on edge. We don't know enough about the species to know if they'll attack or not."

Alyssa considered this briefly and thought back to when she was with the visitors. After a pause of thought, she spoke up in response. "No. It's not that. I see where you're coming from, really, I do. But it's not like that. It's hard to explain because... it's more of a feeling, I can't put it in words very well..." Her eyes trailed away as she tried to pinpoint comprehensible terms. Any better way to put it. "When I was with them, I felt all wrong. Like I was a prized pig being taken to slaughter. They didn't say anything, really, and didn't do anything specific to put that sort of feeling in my head, but still—"

She thought of a specific instance.

"I told them we had a dinner planned, and they seemed really annoyed that we would even go that far. When I told them how to get a hold of me, they were thoughtful about it, but it was almost like they were amused I thought they could need me? When they spoke to each other, it wasn't really speech at all— just tone and sensation. You *are* right about the snake thing, though. That is exactly what I felt like they were. But they weren't garden snakes, they were cobras."

"Maybe it'll seem different tomorrow?" James could only murmur, rubbing her back in a comforting gesture.

"Well, there is another thing, too. I saw the maintenance man, Mario?"

"Mhmm?" He nodded for her to continue.

"I had stopped to tie my shoe and it felt like there was someone behind me. No, actually— I just pretended to tie my shoe specifically to see if whatever I felt behind me was going to come out and scare me. I thought it was just my imagination, but— then Mario came out of nowhere and pointed the same thing out to me. Telling me that someone... or rather, something was behind me— and that it felt unsafe. He didn't like the visitors either, to the point that he walked me home because of how dangerous it felt to him."

"Well I am grateful for him walking you home. I definitely don't like the fact that you felt that way." James got up and reached his arm towards her, silently asking her to put a hand in his.

"I'm worried about tomorrow. For now, I think I just need some rest." She confessed, taking his hand as offered. It curled around it carefully.

"All we can do is make sure we are strong enough to handle it. Let's go to sleep. You definitely need it. Now is not the time to be fatigued."

Together, they went off to bed.

Shortly after he had seen the Captain off, Mario had been heading towards his own house when a noise could be heard around the corner. He knew instantly it was the visitors, or at least felt like he knew. Couldn't exactly say how, but there was an overwhelming sensation that went to the back of his neck and raised all the hairs on his arm– feeling as though he was hunted prey. As if there were a panther in the brush, and he was about to make eye contact– at which point, everything would already be over. They had no reason to be in this area, and Mario didn't know how to get past without being seen. Decidedly, he took out the master key card from his pocket and slid it across what he knew was an empty housing unit.

Outside the door, he could hear them communicating. There wasn't an exact word for the sounds that could be heard. Hissing was involved. Since this room was not in use, Mario had to rely on the darkness to hide him. He crouched down in a corner, and listened intently. He couldn't say why he was so scared. Breathing as quietly as possible, he peered over the lip of the window unit and looked to the hall.

The scene he looked out on was not what he expected at all. These figures were tall– cloaked and hooded that any discernible features were hidden. It looked like there were three of them hovering over the door of one of the residents. As the maintenance man, he came to know quite a few of the people who lived in the residence hall and which room they were in just in case any had needed help. In this specific unit, it was Nancy and Bill's place. They had a son, didn't they? It was Randall, or Randy, or something.

Mario didn't like that these things were sniffing around the door and... it seemed that was quite literally what they were doing. The visitors stood extremely close to the door and sniffed it.

The residents looked weird lately too, but the man thought that it might have been because they had just cleaned up for the visitors. The bike that their kid used wasn't there, and the neon sign and stickers that you'd expect from a boy to decorate were gone.

Without any sort of signal, the door opened. Nancy had an expression on her face that he had never seen. It was utterly blank, with no expression whatsoever. She stepped to the side and allowed the visitors into the house, with the door promptly shut behind them.

Mario took that as his cue, ducking out of the empty home and slipping past with no notice. When he returned to his own unit, the exhaustion hit. And he felt like his skin was crawling. Should he wake up the Captain and tell her? Was there anything to tell? It didn't seem like the man would be getting any sleep tonight. He made sure the chain was on his door, and sat in a chair in front of it.

They wouldn't get him so easily, *that* was one thing he was sure of.

CHAPTER SIXTEEN

J eff had finally decided to open the object that he had kept safe for Justin throughout the years. Having grabbed it from its hiding place and wrapped with a towel, he needed an assured private space to look at its contents.

So it was chosen to be the bathroom— locking the door and starting up the shower at its highest heat, Jeff waited until the room was steamy and its window covered in fog. He sat himself on the floor, undid the towel, then ripped off the foil to reveal a box. All this time, what Justin had been keeping hidden within the box was an old school floppy disk.

This instantly pissed off Jeff. These things could get ruined so fast with even the slightest change that he placed it under the towel and immediately turned off the shower. He didn't know if the information on the disk would even be viable now. But, at least, he knew he could open the files and see if there was any information left.

The rest of the night– most of it, at least, was spent going through his boxes, looking for one thing in particular. His old computer. Jeff didn't know why he had kept it for so long, but now he was glad, as it still had a slot to insert floppy disks into. Even after all these years, it was still in working state after all the times he'd taken it apart and put it together with all the new technology. Around four in the morning, the device was fully up and running. He popped in the disk and opened the file.

A full day and a night's worth of time was spent entirely on the computer, having read through codes and notes and everything in between– and now, Jeff finally understood. The floppy disk was ejected and promptly hidden, and his computer was turned off. It was exhausting, leaving the man to feel strung out at the same time. When he slept, it had been a full fourteen hours until he awoke again. It didn't help Jeff in the slightest, but at least he now had a clear head. He knew what he had to do.

Getting dressed and heading to Earth Point was quicker than expected. His objective was apparent, and he knew that what he was doing was right. With that in mind, Jeff walked up the steps and into the control room. IT was empty during this time of night, but he locked the door behind him to ensure that he wouldn't be interrupted.

It was the process of hacking into the system that took Jeff longer than expected, but they were Justin's codes rather than his, after all. Once broken in, he rooted around the system until finding the personal codes for the Captain. Then, all he had to do was wait, and hope that Alyssa wasn't a deep sleeper. He waited. And waited. He even pinged her on her computer every few minutes.

Finally, he received a reply.

<Hello?> The typed response from the Captain read.

<Are you alone?> Jeff typed.

<Who is this?>

<A friend.>

<I'm alone.>

<I need to talk to you face to face. But not now. On a secure line.>

<When? How? And how are you talking to me right now? We have been radio silence with Earth Point.>

<I know. I'm sorry. Tomorrow night. Three AM. Main comms room for you. I'll set it up. Make sure you are alone.>

<Why should I trust you?>

<You shouldn't. But I'm the best hope you have.>

<For what?>

<To get out of there.>

<What do you mean?>

<Tomorrow night. Alone.> Jeff quickly disconnected the line and erased the content. Any evidence that the conversation had taken place.

Having unlocked the door, he made his way towards his desk, set on returning to work as normally as he could present himself as. When he got to his desk, unfortunately, he wasn't alone. Alex was behind him when he spoke. "We need to talk."

Jeff needed coffee. Of course he needed coffee. That was the one thing he needed to combat the fact that he was shaking from head to toe thinking about the prospect of having gotten caught. Though he had never done anything wrong his whole life, just now he had snuck into the communications room and sent out unauthorized contact. While the higher ups won't know what was sent, as he stood, the man looked incredibly guilty.

"Hey Alex... what about?" He said in an attempt to be nonchalant, rubbing his eyes and then walking deliberately to the coffee maker.

"I need to know something."

Oh god. Jeff knew that the likelihood of him being able to lie outright would be impossible. Going around the truth? Fine. But straight up lying, no way.

"What's that—" It wasn't quite a question.

"Will you stop and look at me?" Alex demanded, grabbing Jeff by the shoulders and turning him around.

"What?" The expression on his face was exasperated.

"I need to know if you're okay to be here." When his gaze met with his coworkers, he could see the concern that filled his eyes. Even Alex's voice was laced with worry.

Sighing in relief, Jeff's lips almost curled into a smile but he held himself back. This was a question he could answer. "Look, man. My best friend is gone. I'm hurting. I'm not going to front about that. But I'm okay. Being at work, having some normalcy— it's what I need right now." Jeff exhaled, and though his hands had been trembling slightly, they placed themselves onto Alex's shoulders with a pat. "I just need some coffee and to go through some codes. I'm not okay. But I'm not broken."

"Do you need some time off?"

"Did you hear me? I need to get back to work. I'm okay. I just need to read some codes. I'll be okay. I just need to work." He snapped and repeated a few words that he had already said, things that Alex already knew, but at least it was enough for the other to back off.

"Okay, well... look. I'm here for you. Just let me know what I can do to help you."

"You can tell me what happened to Justin. Find that out and then I will be fine. Some form of closure, y'know?"

"Yeah. Okay. I can do that."

And then he was alone again. Jeff considered making another pot of coffee to sit and think about what needed to be done before tomorrow. What he said to Alex wasn't necessarily a lie. He was going to do exactly what he said he was; work on codes, and figure out how to save an entire space port from what he learned of. Secretly.

There were a lot of codes to get through. Feeling like one of those old-school hackers you'd see in the movies, he went through every line of code he needed and wrote out his notes on a legal pad. Jeff had about eight pages worth of notes, front and back filled by the time his shift ended. He took his notepad with him and left for the night.

There wasn't going to be any attempt at getting sleep tonight, of course, though he had decided to comb over the codes written down and figure out new commands to try. As a programmer—hacker, in this case- he knew lots of ways to get through things, firewalls and the like. This time, he contemplated looking into the codes of others. Personal computers and breaking into them had never been a problem before, and he doubted that it would now... but this had to be done carefully. Anything could be traced, especially suspicious activity. Jeff decided that this wouldn't be attended to until the last possible second. He needed to be involved for as long as he could. Save the more dangerous things for last.

Charts were made in his head for a plan of action— even going as far as making back up plans to his back up plans. Jeff went to fetch a duffel back and packed everything he might need for a quick departure. He stopped in place upon closing the bag, realizing a vital component for his plans: he needed to find a way to get mobile. A simple solution; he'd buy a car tomorrow, have it delivered, and then he would be set. Now... he just needed to figure out what to say on his call tomorrow.

CHAPTER SEVENTEEN

There had been a lot of thoughts spinning through Alyssa's head after conversing with the stranger from Earth Point. After they had been struggling to have any sort of contact with them, these messages felt almost cryptic. But the foremost thought above all else was that she was tired. Oh, so tired. She'd never felt this level of fatigue before– she needed sleep, and she was going to get it as soon as she collapsed into bed.

Alyssa slept like the dead, arms wrapped around her husband and deep in dreams.

"You got yourself into this mess! Get yourself out! Remember Alyssa, for every mistake you get yourself into, there are at least two ways out. Find one! Find one now!"

She was in the ring with one of her father's soldiers. He had been teaching her how to defend herself for years, now. There were a lot of lessons taught in the ring, and this was just one of them. Some of the others included things like, "Most fights will end up on the ground. Think of the way to get out of it when you're there." and "If you don't protect yourself first, you can't protect the people around you."

The wake up was slow. Alyssa gradually blinked away the past memories from her slumber, having laid in bed with James longer than she should have. When her husband got up to shower, she remained beneath the sheets, thinking about her father. He wasn't the best father in the world, nor did they have the best relationship, but he did what he could.

And so she did. Joining the military like her father, getting married like her father– the whole getting married and having a kid, just like her father. She was her father's daughter, and somehow, she needed to be just like her father again, today. Alyssa felt something amiss in the air. Something was wrong.

Coffee with James was nice, though. Once she was finally able to get out of bed, Alyssa confided in her husband about the dream and the memories flooding into her. Eventually, she admitted to the disturbance that her intuition had been pointing her towards. The air seemed to feel... static. There was something off and she wasn't quite sure how to put it.

"Yeah, babe, there is a change in the winds today." James could only nod in response.

"A change in the winds." She repeated, gaze distant.

"Well, look. I got to go." The man shuffled around, looking for his bag, while Alyssa silently watched in contemplated silence.

It didn't take long for her to speak up, albeit hesitant at first. "Hey. I need to ask something. And it is terrible, and I know it's crazy, but–"

"Hey. stop rambling. Whatever it is, just tell me." The bag that James held over his shoulder was placed down onto the ground while he looked at her from across the room.

"...Don't go to work today. Keep Steven out of school. Stay out of sight. I don't know what is going on, but I can't worry about work and you." She sighed, fretting, arms crossed and pulled into herself.

"Oh hun, I don't know. I can't just–"

"*No*. Listen to me. I can't explain it. I just know something is wrong– and I need to be able to count on you not being around if–" She almost corrected herself by saying 'when', but held the word back. "– something were to happen today. Can you just do that for me?"

"Yeah. Sure. I can do that." Eyes lingered on her for a long time, confused and startled– but James said nothing else, before he grabbed his back from where it had dropped, and headed back to the room to put his stuff away.

Now that she could count on her family being out of harm's way and within their dwellings, Alyssa could feel as though she could walk out confidently and try to save... whatever was going to go wrong today. There had already been a lot going on that a day like today would be the perfect opportunity for shit to hit the fan. Walking into the main control room, she was met with her crew boisterously shouting back and forth. There were papers on the floor, and nobody could actually be heard with their voices overlapping one another.

In the middle of the chaos, Alyssa saw who she needed.

"Mary!" Alyssa shouted towards her as she walked over to the communications table. Flipping the comm speaker of the room on, the Captain yelled out.

"*Everyone! Stop!*" All at once, the room went silent. All eyes were on her.

Shutting off the speaker so that she could speak normally, the woman raised her voice, "Seriously? What is going on? I need Mary to give me an update, and the rest of you to get your stuff together and wait. Calmly," she emphasized that single word, "...for the next move."

As the shuffling of individuals regaining their documents became the only sound heard within the room, Mary walked up at that point. A bit haphazardly did she begin to rattle off what was going on. The unfamiliarity of her crewmates' anxiety made her feel... much worse.

"All access points to Earth Point have been completely broken. There is no way to fix it on our end. Everything that is our system is on the fritz. The entire place. There is nothing that can be fixed on our end. We can't..." Her sigh was long and drawn out, tears swelling in her eyes. "No one knows what to do. We are completely locked out of Space Port. Nothing is working." All connections to any family, anything on Earth– gone.

"Okay. Is there anything else?"

"Um. No. Well, yes– the visitors–"

"What about them." It wasn't a question. It was a demand. Insistence to know what else was going wrong on her ship, deadpanned in her apprehension.

"They have demanded that everyone meet in the main dining hall. Most everyone had begun heading in that direction and none of the crew had been able to stop them. More than that, we have pinpointed about twenty-three missing people that are unable to be found within the ship grounds. Not a single soul knows anything of these individuals– and our supposedly trusted visitors continue to act and command without your permission. All within ten minutes... pardon but Captain, I am freaking out!" The words could only spill and spill forward like a dam that had broken loose while Mary paced throughout her ramble.

"Okay. Shit. Okay. There are a few... That's a... okay." Alyssa ran her hands over her head as though she were bald. "Wait.

Tell me again. In order. Short things. Uh– Words, I mean. Give me words." She was breathing deep and trying to focus on the problems in order.

"Yes ma'am–" Mary inhaled sharply and began again, a bit slower and more thoroughly this time. "Um... in order, it goes like this: The visitors feel ominous. Somehow, they are gathering everyone into the cafeteria for some reason, and I'm not sure how they were able to talk to the masses without going through us. We only realized because when on his way here, Mike had overheard a few people already on the way there– As soon as he arrived here and we all began our shifts, there was an immediate error code indicating a system wide shutdown... Since then, everyone has been offline Everything, everywhere. We have zero control over any of our systems. And when we ran counts of our total occupants, as that was a separate program– twenty-three individuals were not logged." As her frantic breathing began to subside, the woman let out a soft exhale. "So. There it is, Captain."

"Alright. First– I think we should try to figure out what is going on in the mess halls. Send Mike and someone else there. Tell them to hang in the back. I don't want them noticed, and I don't want them really involved. I'll try to get a handle on the missing people. I may have a few ideas on where to search for them. Everyone else needs to divert all energies into getting things up and running here. Oh– put a webcam on Mike so we can keep track of him and maybe see what's going on without him having to speak. It doesn't have to link to anything so there will be nothing to stop it. Just a one way camera where we can see what's going on. You're going to be second, here, while I check what I need to check." Alyssa stood up and put her hands on

Mary's shoulders. "It's going to be alright. One at a time, okay? Mike, camera, then all your focus is here. Have Billy watch what happens with Mike."

"I'm direct, Captain." And with that, Mary walked out and the room went into go-mode. Since the whole conversation happened in the main room for everyone to overhear, the crew now knew the situation and had a plan to follow.

Alyssa headed towards the medical bay. There was a list of everyone, every day, and that was the list she was going to use to see who was missing and go from there. And though she was going to find those people, she was going to start with the corridor she had given the visitors. But– wait. She had an idea.

Doubling back, she went back to the main comm room and picked up a set of old school walkie talkies.

"Mary, heads up!" Tossing a walkie talkie to Mary, she started barking orders out. "Everything is down, but walkies have a range that should encompass this whole craft. I'll be on two."

"You're brilliant!" Mary shouted at her, turning on the walkie.

Alyssa keyed up on the walkie. "Mary, you on?"

"Yes, Captain."

"Did all the visitors go to the mess hall?"

"Yes, I counted them on the video feed. What are you going to do, Captain?"

"I'm going to their corridor and I'm going to see if any of our missing people are there. Radio me if anything changes or if they break to come back."

"Direct, Captain."

Now that the plan was in motion, Alyssa couldn't stop the pacing of her feet, or her thoughts. She couldn't stop thinking about Steven, and she didn't want to get started with worries of her husband either. It was already noon. She was losing time. Picking up the pace to a sprint, she made her way to the next hallway. One more turn and she would be at her destination.

The hall seemed cold. Rather, the air was extremely cold. Frigid to the point that she could see her breath. This fog seemed to billow around her at eye-height as she walked quickly around the corner and into the first room. An empty, unlocked room with nothing in it. The next three rooms were the same. As for the last three rooms... they were the jackpot in the worst sort of way.

There they were, the twenty-three missing individuals that had all been lost over the course of days. Having opened the doors of the last three rooms, it was the sight that greeted Alyssa. Twenty-three people, still like mannequins.

Just. Standing.

"Hey! Guys!" She hissed, waving her hands frantically to gather their attention— yet not one person was responding. Even as she stood in front of them, in direct line of sight, they acted as though she weren't there.

They were staring. Vacantly. At nothing. Alyssa touched one. He didn't move. Then she grabbed him and tried to get in his line of sight, trying to talk him into responding, but... nothing.

"Come on, look at me! What did they do?"

No one had spoken to her for a while— no matter what she did. She wondered why, before looking down and realizing she still had the radio in her back pocket.

"Mary." She whispered into the device.

"Yes, Captain?"

"I found them. I... I don't know what's wrong with them. I can't get them to talk to me. They are just—" She didn't know what to say. "Standing there. Vacant eyes. They don't see me. Not really."

"Captain? They are leaving the mess hall. You need to get out of there."

"I can't leave them, Mary."

"You're going to have to. We don't know what's going on yet. You can't be alone with all of them. Get out and figure out how to help them."

"I... have to leave them." After a long pause, she began murmuring to herself, walking out of the room. She repeated it over and over as she left. "I have to leave them, to save them. I have to leave them to save them. I have to..."

"Captain, Mike came back. With a visitor. I'll let you know what is going on once I find out, but you should get back here. Now."

CHAPTER EIGHTEEN

J eff was having a bad day. Too many bad days. He had been
pretending to manage such days well– starting to get good at
it too. No one suspected that on the inside, he was freaking out a
little. Just another day to pass through in order to get his affairs
in order.

The first item on his task list was vehicles. He called Alex first
thing.

"Good morning!" Jeff chirped, then realizing that he sounded
way too chipper, added, "Asshole."

"Good morning to you too, Jeff. What do you want this
morning? Because you only call me in the morning if you need
me. What can I do for you?"

"I need a vehicle." The request was made as casually as
possible.

"You... *need* a vehicle? For what?" Alex asked.

"I can't keep bumming rides off of everyone. And eventually I
need to go see my family, and Justin's family. I will need to be able
to go places!"

There was a silence on the other side of the phone. Too long of
a pause that Jeff had no idea what to make of it.

"Jeff–" Came the voice on the end.

"Alex, come on. You know I can't do this. I have to have a vehicle. I would feel a lot better knowing I had my own ride... today." Though the urgency bubbled in his chest, Jeff tried to not sound too much.

He heard a sigh, and Jeff knew by the sound that he had won this battle.

Alex had given in. "I'll tell you what, Jeff. Take Justin's truck. No one is using it. We can't give it to the family, yet. So until we find something more suitable for you, I can let you take that."

"You can let me take it." Jeff repeated, a tinge of anger in his tone. "You can *let* me take it? Like it's yours? What the hell?"

"Jeff, take it or leave it. Justin wrote it to me in a standing will when he moved on sight. So yeah. I can let you have it." It seemed as though every word that came out of Jeff's side only brought his friend down. Alex sounded exhausted, to the point that his voice seemed strained.

Instantly, Jeff tried to lighten himself up in contrast. The man on the phone sounded bad. Worse than he remembered. "Alex... are you okay? What's going on?"

"Nothing. I'll drop off the keys when I take a shower and have some coffee." Before more could be said, Alex hung up the phone.

Jeff couldn't help but smile, brushing off the other's mood today. Something worked out. Something had worked out well for him. He didn't have to lie that much, and he didn't have to try so hard. It was one thing he didn't have to worry about. Now, he would pack.

The duffel bag he needed could be found in the far back of his closet. Jeff filled it to the brim with anything that he could possibly need; clothes, toothbrush, toothpastes, soap bars, and even some power bars. He placed the bag and his most comfortable pair of shoes by the door so that he'd have them ready to go. He needed to be ready to go, when the time came.

Going into the kitchen, he realized he smelled. He smelled *bad*. Time for a shower. Things were going quick *and* slow today. He needed time to think, but he also needed to get things ready. There was never a good time to think and do things that required thought. One of the things, or the thoughts would eventually get messy.

The shower was neither hot nor cold. Even if it was, he wouldn't have really noticed. Jeff was thinking about what he found on the floppy drive. For one, he had no idea how Justin managed to get all of that information on the floppy drive. For another, he had known about this for so long. They had never talked about it. Why had Justin never told him?

The program was all numbers. It had to be broken down by code. Only a handful of people in the world would have been able to do it. Jeff was surprised he was able to. Guess all those code games Justin had made for Jeff came to good use, finally.

It was a complete breakdown of what was actually happening. How the war started, how it ended, and why the space port was so important to the world now.

Jeff knew he had to tell Alyssa. But... how to fit this all in one conversation? Maybe he didn't need to. All that needed to be done was to give Alyssa a fighting chance. Saving them all wasn't possible. It was too late for that. Although maybe, just maybe, he could help in some small way.

The water began to run noticeably cold amidst his thoughts. He got out of the shower and thought of his own next step. The only next step that would matter.

Alyssa was point break running. She had to get out of the hallway. The entire day was a blur. She was trying to get the families together. But it seemed that the visitors had done something to the entire crew. No one was where they were supposed to be. It was a ghost town, literally.

People were wandering the halls aimlessly. No one was listening to their names. No one was reacting at all. She slapped a few, threw a few punches even, and still the people around her had just... stood there. She couldn't reach them. And she noticed something else, too. There were no kids. All of the kids that had been on the space port were just gone. Poof. Vanished. How was that even possible?

She had been trying to find her family for hours now. While looking around each corner, making sure she wouldn't see any of the visitors. It was almost time to meet though. So she threw caution to the wind and ran. She ran with all her might through all the service hallways and through rooms she knew were totally empty. All to get to the one person on Earth Point that could possibly tell her what was going on.

Jeff was at work again. He had just a few more minutes to try and get the room to himself so he could talk to Alyssa. The last person was dawdling, though. Jeff almost wanted to help the old hag get her things together to hurry her along. But he had to be cool, calm and collected. He nodded at her when she passed.

"Have a good night." She mumbled, and he nodded back at her.

Logging into the system, he paused. Looking over his shoulder, he made a rash decision to lock the doors so no one could come in. If someone tried to get inside, he would at least have a heads up and be able to log off and look innocent. Maybe feign like he was asleep and didn't know the doors were locked.

Entering the code that he found on the floppy, he accessed the remote access to Space Point and waited for Alyssa to accept the meeting icon.

One minute later there was a beep and Alyssa was on the screen before him.

"What is going on? Who are you? What is happening here?" Alyssa was frantic.

"Look, I can explain as much as I can, but I need to tell you some things first. We don't have a lot of time. Let me tell you everything and then you can ask things later."

She nodded and he continued.

"It was all a ploy. The war, One World United, everything. It was a ploy that the aliens and One World created to ensure Earth's survival. The land in space— it's owned. By the aliens. Your visitors are the landlords. You guys are payment. Do you understand what I am telling you? Your entire ship has been sold to the aliens as payment for more space to build ports. You've been sold. Like slaves. I don't know what happens next, but that is why there were so many health checks. They were making sure you guys were viable." He took a breath.

"What– what do I do?" Alyssa breathed.

"Look. They got rid of your emergency space craft. I don't know what to do now. I don't know how far this goes or how to get you out."

"There's no ship." She inhaled slowly.

"Not one of ours."

When she looked up, despite her initial behavior, there was a fierce look in her eyes. It was at that moment that Jeff realized how she became captain. The tenacity that Alyssa held within her was one that would make legions follow her into battle– all with a single glance. Even he was willing to do anything for her at this point.

"While not a ship of ours, it is a ship nonetheless." She was looking around her as if trying to calculate in her head what this actually meant.

"Can you even fly an alien spacecraft?" Jeff sputtered out.

"I was trained to fly anything I was put into. It shouldn't matter what it is. I think I can do it. But– where do I go?"

"My family's farm. There is a land big enough– no, they would track you there. You'd have to land anywhere you could and then travel by foot there. Can you get provisions?"

"I have an emergency backpack with enough food for three people for a week. James also has one. That can keep us for three weeks if we ration."

With some plan in place, Jeff gave her the coordinates to his family's estate and knew that when he left here, he would have to

ONE WORLD UNITED

haul ass to his family's land and let them know what was going on, and that he had just brought the war home with him.

"What if I can't find my family?" Alyssa blurted out, her voice cracking as she spoke.

"I don't know. I don't know how I can help. I wouldn't have even been this helpful had it not been for–"

The reply was broken up by someone trying to get into the door. Jeff froze.

"Disconnect us. I'll find you when I'm back on Earth." Alyssa whispered quietly.

"Good luck." Jeff stammered before ending the call. He ended the call, deleted the file and went to his desk and grabbed his stuff.

He opened the door and the next shift was standing there, looking at him.

"Hey sorry, I left my bag and I think I turned the lock for some reason when I turned back to get it. I'm too tired. I'm gonna crash."

"Been there, man." One of the guys said to him.

Jeff left the building as quickly as he could, not seeing anyone on the way out, and he got in his truck and started to drive. And he didn't stop. The great thing about this base was no one cared who was leaving, only who was coming in.

211

CHAPTER NINETEEN

Alyssa's head was buzzing, but like any other time that she had something pressing, she remained calm. The cacophony of a thousand anxious thoughts didn't matter. That was one of the things her father had passed down to her; crisis didn't freeze her. It didn't freak her out, not anymore. When a catastrophe were to occur, the discord calmed her instead.

There was a checklist in her head. A set list of things she had to do. The first thing being to get to her family. She didn't think they would be in the quarters, but she had to go there anyways for her and Justin's pack. That was decidedly her first destination, looking for any of her family while she went.

None of them could be found on the cameras. This was both a good thing, and a bad thing. If Alyssa couldn't find them, then maybe the aliens couldn't either. She knew the layout of the ship far better than they did, and this fact was also beneficial. The layout for the rest of the cameras came easy to her, and if she wished, Alyssa could avoid being seen.

She was back to running. Through the service corridor, down the medic hallways– wait. What was that?

Looking around the corner as she had been doing, to ensure no one of the alien persuasion was around the bend, she saw a line of people– her people, heading into the room. One of the aliens stood there, seeing them in one at a time. The leader had to be inside. Alyssa could just make out what they were doing; their leader cutting a piece of himself, then placing it onto eyes of the crew. They were already in a trance-like state, but once that silvery piece touched their skin, dull eyes turned cold and pale. Blues were turning into an icy shade, with brown much like pale sand as vibrant greens were now muted. There was a mark on the side of their throats. It was like a silver birthmark– the exact silver of the alien's face. Mirror-like colors that had scared her to the bone.

She recalls the markings on the people she had found in those rooms just yesterday. This is what happened to them, this transformation. The anger and the fear surged within her. An urgent thought. Where was James and Steven?

She bypassed that hallway and scurried down another– which also had to be avoided entirely, since a flock of visitors were taking up space down that direction. Hurriedly skipping over the new hallway altogether, Alyssa made her way down yet another one. Frustratingly enough, she had to find some other alternate route as she found that visitors were making their way through there as well. It was one detour after the next, over and over, until she finally made it to her personal quarters and slid inside.

"James!" A sharp yet quiet hiss into the room.

Alyssa wasn't betting on an answer, but she remained waiting for a reply. Next room to check was Steven's, stepping through the open doors and looking around. There wasn't anything needed from here, but the confirmation that her son was not here had been all she needed to make sure.

Then, it was the shared room between her and James. An immediate stride towards the closet, backpacks nestled on the floor that she needed. They were filled with provisions that would be needed when they got to Earth, after all. She shouldered her pack on her right shoulder, with James's pack on her left. Moving quietly, she snuck out of the quarters and looked both ways.

"Psst."

Looking around wildly for the source, Alyssa flattened her back to the wall and breathed slowly. Trying to see if she could hear the sound again.

"Pssst." A voice came again.

"Hello?" A low whisper beneath her breath, looking around again. Nothing of note. It was deserted, but Alyssa was certain she heard something.

"...hello?" She murmured again.

"Up here." The sound came from above.

Turning her gaze up slowly, eyes were met with Mario standing atop the ceiling rafters. "Mario!" A shocked whisper, surprise in seeing the man.

"Down the hall there is my ladder, Captain. You can use that to get on the low beams. Just bring the ladder with us so we can't be followed."

"I have to find my family." Her head shook slowly with rejection, turning down the lifeline that dangled in front of her. "I can't leave them."

"They are with me, ma'am. Just in another room. I got Steven up after some people headed to the cafeteria, and we got to your husband together. Hurry, please." He signaled with a hand for her to move on.

With a slow nod, and a soft sigh of relief at hearing her family's status, she jogged her way down the corridor towards where Mario had pointed. The ladder sat there, just as he said. Climbing up with haste, Mario moved past her to wrap rope around a step, pulling it up onto the rafters with them. He carried it with him across the beams, urging Alyssa forward with a free hand.

"Where are we going?" She couldn't help her gaze scanning their surroundings.

"Hide. We hide, *now.*" He hissed, putting a hand on the small of her back to lead her down into the next room.

Crossing a narrow archway, they went into another room, the cafeteria, and there she saw Steven and James. She wanted to call out to them, but knew they had to be quiet. Thankfully, both Steven and James had the same thought. Steven waved, but didn't say a word. Just waited for her to cross to them.

As soon as Alyssa could catch up safely, she wrapped her arms around them more tightly than she ever had before. Tears strung her eyes. She had her family. Everything was going to be okay. They could get through anything, as long as they had each other. They were going to make it out of this.

"Alyssa, what the hell is going on?" James whispered to her.

"Too much to explain. But we *need* to get out of here. Is there anyone else that has not been infected by the aliens?"

"Not that I have seen." Mario replied.

"Everyone. I think they got everyone." James breathed out.

Steven piped up, "Not us though."

"Shhh." Alyssa knelt down, looking around as though the words would carry and give away their position. For now, it didn't seem to. "...you're right, though. Not us. But we have to be

ninja quiet and stealthy now. I have a plan." She said to her son,
then looked at Mario and James in turn.

They sat together in the corner rafters of the cafeteria making
a plan to get to the alien ship. Along the way, they would look for
survivors.

There were people milling about. But, the odd thing was...
it was silent. Everything was so quiet. No one was talking, not
one person. The sounds of the ship that Alyssa had become so
accustomed to were muted out by the zombie-like persons that
used her crew. It sent shivers down her spine to think about what
would eventually happen to them. She had no idea what to do—
as far as she could see, they all had the same vacant, glassy eyes
that the people in the rooms had. They droned the same, mulled
about... wrong.

Could she do anything at all? Save them? It... *no*, she couldn't
do anything for them, not with her family on the line.

Their hope of finding any survivors was crushed.

And with that, arose another issue with this plan: they could
only get so far in the rafters before they had to get down.
They'd be with the masses down there. At least, they'd be *seen*
down there. The ship, as far as Mario knew, was almost entirely
surveilled by two aliens. There would have to be a diversion, or
a fight. There was no way to know how to fight the aliens. The
visitors did this with as little droplets of self-made 'quicksilver'.
How would you even begin to combat this?

"Mom, I have an idea." Steven whispered.

James shook his head at his son, though, and the grown-ups
kept trying to work through the problems at hand.

Mario eventually sighed, "I don't know what to do."

"Mom. *Please*, listen. I have an idea."

Even if Alyssa was side-tracked, she stopped and looked at the boy. No one else had any feasible solutions. It was all too confusing and terrifying, especially for a child Steven's age. But he remained enthusiastic despite it all– even when the adults surrounding him began to crumble under the pressure. He was so brave.

"Tell me."

He went into his backpack that he had on. She hadn't noticed he was wearing it. "We use this." Grinning, he took out his baseball.

James smiled and lifted his son into a hug. "I'm sorry, kid. You're brilliant. Yes, that will work. I can't believe how amazing you are."

After some discussion, their revised plan was set into motion, and they continued through the long trek above the ship, as far as they could go.

There was still no noise below. Every footfall made Alyssa wince. She felt like it was amplified by the rafters. Any minute now someone was going to look up and they were going to be found out. Then the chances of them getting through the ship safely would be next to nothing.

Suddenly, a loud screech sounded out. The P.A. system, Alyssa thought. It happened when the mic was too close to the speaker. Someone was in her chair.

Everything went dark.

There was a button to lights-out the entire ship. It was used seldomly, but Alyssa knew *they* had pushed it. Mario, who was up ahead, stopped and crouched down. James, Alyssa and Steven did the same.

There was an ear-splitting voice that echoed out through the main speaker system, *commanding*.

"We are missing four humans." The *s*'s were drawn out and sent shivers down their spines. "We must have them. If we do not, all of you will suffer. Find them. Bring them to me."

The entire shipful of people were now running around in careless circles and aimless lines, looking for the people that were not like themselves. Four people who were not yet dosed up on the aliens being. Somehow, each person who had been affected knew each other. Like they could collectively sense the presence of the alien inside– like a hivemind. They searched every room and corridor for the small group in the rafters.

"This complicates things." Mario whispered back.

They didn't have much longer before they would have to be on foot, on the ground, with a city sized population of people looking for them. The hope was that they could get close enough to the ship that no one would be there, and then they could create a diversion, one long enough to steal away on the ship.

The reality was, that there was fighting and screaming and *chaos*.

They had indeed found their way down the ladder and two hallways away from the ship, but one woman saw them and screeched while pointing at them. It was *deafening*. Alyssa stared at who used to be her shipmate. The noise that came out of her mouth now was not human. The sound terrified Alyssa– *this person was no longer human,* no, not at all, how could she be? There was no way any person could even physically make *that* sound, they just couldn't.

Even if only for a moment, Alyssa froze thinking about the state of this poor woman. It was horrifying, she couldn't suspend her disbelief– it drew on long enough to have Mario doubling back and grabbing her.

"We *have* to keep going, Captain." He yelled at her.

Everything happened so fast after that. People came rushing in from everywhere. They were all howling rabidly, so loudly that Steven had to cover his ears while they ran. Shrieks from every direction– anywhere they could turn. It was overwhelming, so much more than any of them had ever heard or prepared for.

There was no end to the group running into people. Mario swung the ladder this way and that, pretty soon making a barrier between the horde and the family.

Oddly, Alyssa noted that there were no smells from any of these 'people'.

No one's breath clung, even though she could literally feel them on her neck– really, everywhere around her, they were swarming. There should have been the scent of the masses running and joining together, some foul and some sweet, the way humans were. They should've especially been off, controlled this way or running as they did.

Even if the ladder struck a few, she couldn't even pick up any blood in the air, only that of the metal.

Mario and James rushed the doors to the ship and doubled back, making a stand against the rising crowd while Alyssa tried to get the doors open. This 'hivemind' was pushing up against the ladder to where it was pressing against them, their only real shield. She had to use her back to brace up against James to try and get to the keypad.

They were all gasping for breath at this point. If Mario and James let go of the ladder, they would be smashed against the wall until they stopped breathing. Alyssa put her knee up against the wall for better leverage against James and grabbed Steven so he could huddle down in the space between his mother and the wall.

At last, she unlocked the damn thing, but there were so many people pushing up against them now, she still couldn't pry her way in. Turning to her husband, she tried to assess the situation and figure out how to open the door. Her son was desperately trying to get into his backpack again. Alyssa was pushing him behind her.

This was it. They couldn't make it. They had fought so hard and she wouldn't be able to open the door to save her family.

Steven had his entire arm inside the backpack. He was shaking uncontrollably as he searched for *something*.

Finally, he cried in relief– the little boy pulled out a flare gun.

"What are you doing with that?!" Alyssa gasped in horror. She was fumbling just trying to get her hands on it, but she couldn't possibly manage it between everything else.

"Dad!" Steven shouted, and threw the gun to his father instead.

On instinct, James turned, caught the gun, and turned and fired it into the crowd.

The shock of the eruption, bright and hot, did exactly what it should have done: the group reeled. Some shrieked in a derangement of visceral horror on too-raw throats.

The boy gave his mom his best attempt at a smile. "I found it in the service emergency boxes. I thought that I would just save it in our bags just in case we needed it somehow."

The entryway was able to be forced only wide enough for the four to squeeze through. The door shut behind them and immediately the throng came rushing towards it like a wave crashing along the rocks.

There was no time. No time to think, no time to breathe. They were running flat out to the ship now. A loud chorus resounded behind them – *banshees from nightmares* – and the doors were about to give away.

Alyssa knew what she had to do. She moved against the path they were running, and to the wall where she could override the system and initiate a lock down sequence inside the hull. This would enable them to open the bay doors when they needed, but ensure her crew, however compromised, would be safe from the pull of space, once the doors were opened.

"Get in the ship!" She yelled as she went to the lock screen on the wall and began entering in the sequences needed.

Her husband, wide-eyed, *looked at something she hadn't seen yet,* shouted back without thinking: "The other aliens!"

"Dad!" Steven screamed and grabbed his bat that was sticking out of his pack and threw it to his father.

"Captain, we *must* hurry!" Mario's eyes were set past her from where they'd been, to the crowd who were almost through the doors.

She called back: "Two seconds!"

"*Mom!*"

Mario grabbed Steven and hauled him towards the ship. James was hitting the aliens with the bat, and overcoming them. And Alyssa—

"*Yes!*"

With the code finally finished, metal doors that had been firmly shut now burst open, and Alyssa was already well on her way towards the alien ship.

ONE WORLD UNITED

Once in, there was an immediate jump into the pilot seat. Alyssa shouted at her husband to buckle Steven in while she messed with the controls. There was nothing she could understand on the panels, but all ships had a similar foundation.

Her fingers followed a sequence that seemed most used– what she presumed was the startup, and managed to get the ship to activate. More keys were pressed and the bay doors started to open.

"Uh– we have a problem." She hollered behind her as hands began to struggle, cursing under her breath.

"What?" James shouted back to her.

"The ship is on autopilot!" Alyssa became more frantic, trying to override a system that she knew nothing about. Hasty movements that pressed key after key eventually dwindled in defeat, letting herself fall back into the seat. "I– I can't shut it off!"

The flashing lights of the space port could be seen just at the edge of the captain's vision, while the endless night sky around her only seemed to become more enveloping. There was nothing she could do.

"Where are we going?" Steven cried out.

Alyssa shook her head, on the verge of tears, struggling to find an answer. "I don't *know*–!"

There came a pause.

Life... *abstracted.*

It was the only feasible term for it– an event no human should survive, the brightest of her crew would have told her, she was so sure of it. This was Sci-fi that played out only in the movies, phenomena beyond science by decades, not even captured in the most grandeur theoretical models. Many ruled it impossible, it did not make sense; hundreds of years could not be condensed into fragments of seconds inconsequentially.

Humankind could not race the speed of light.

This concept escaped the fundamentals she relied on most as a captain: control and sensibility.

But her brightest were drones bred and sold like cattle, and they would never know it. She had no crew left to call herself captain of, nor the ship she had been assigned, only what *they* brought to *her*. This escape had been from the very thing she fought so much of her life for, a so-called beacon of hope created from the ashes of a broken world, and... it was taken away before she could give a name to the one taking it, *effortlessly*. They could whisk a hand, and the people she cared for and respected were diminished to husks, alien playthings.

And what could she do but run...? Alyssa was just as human.

The four of them alone stood to face the impossible, and they were helpless to stop it.

Thoughts spiraled on, down and down, caught up in the endless moment she now named 'home', her hands still clinging to foreign modules. Like she could change it after all if she really tried. Like she could still be the hero everyone told her she was. But, the reality was that she couldn't move. There was no changing the circumstances; it remained absolute. The autopilot, the fear in her son's eyes, the sellout, and... the man who tried to warn her before it all... how unfair the war had been for it to be another lie. How many people had really died for this farce?

Were those the last things she would ever think about? How much time did she actually have like this?

Aside from her regrets, that moment was silent.

The cacophony stopped. It all did, all at once, even if it was devouring them whole in the same second. They were locked within a snapshot, a single photo– within a frozen frame in time. The screaming outside, the frenzied wails and slams of former passengers... they were quiet again, still as she was. Pacified. Kind as they were once. There was no breathing, *no crying, no scent, no movement...* Respite. Alyssa almost let out a breath she didn't know she was holding. She wasn't sure how.

It was like this she realized her time was up.

When those hands suddenly felt very far away, things were recognizably changed, the world as they knew it dissolved around them.

The ship engaged hyperdrive.

One World United will return, divided.

Printed in the USA
CPSIA information can be obtained
at www.ICGtesting.com
CBHW030659300324
6000CB00005B/21